European Consortium for
Mathematics in Industry 10

H. W. Engl / J. McLaughlin (Eds.)

Proceedings of the Conference
Inverse Problems and Optimal Design
in Industry

European Consortium for Mathematics in Industry

Edited by

Leif Arkaryd, Göteborg
Heinz Engl, Linz
Antonio Fasano, Firenze
Robert M. M. Mattheij, Eindhoven
Pekka Neittaanmäki, Jyväskylä
Helmut Neunzert, Kaiserslautern

ECMI Vol. 10

Within Europe a number of academic groups have accepted their responsibility towards European industry and have proposed to found a European Consortium for Mathematics in Industry (ECMI) as an expression of this responsibility.

One of the activities of ECMI is the publication of books, which reflect its general philosophy; the texts of the series will help in promoting the use of mathematics in industry and in educating mathematicians for industry. They will consider different fields of applications, present casestudies, introduce new mathematical concepts in their relation to practical applications. They shall also represent the variety of the European mathematical traditions, for example practical asymptotics and differential equations in Britian, sophisticated numerical analysis from France, powerful computation in Germany, novel discrete mathematics in Holland, elegant real analysis from Italy. They will demonstrate that all these branches of mathematics are applicable to real problems, and industry and universities in any country can clearly benefit from the skills of the complete range of European applied mathematics.

Proceedings of the Conference

Inverse Problems
and Optimal Design
in Industry

July 8–10, 1993 Philadelphia, Pa. USA

Edited by

Prof. Dr. Heinz W. Engl
Chair for Industrial Mathematics
Johannes-Kepler-Universität, Linz, Austria

Prof. Dr. Joyce McLaughlin
Ford Foundation Professor of Mathematics
Rensselaer Polytechnic Institute, Troy, NY. USA

B. G. Teubner Stuttgart 1994

Die Deutsche Bibliothek – CIP-Einheitsaufnahme

Conference Inverse Problems and Optimal Design in Industry
<1993, Philadelphia, Pa.>:
Proceedings of the Conference Inverse Problems and Optimal Design
in Industry : July 8–10, 1993, Philadelphia, Pa. USA /
ed. by Heinz W. Engl ; Joyce McLaughlin.
Stuttgart : Teubner, 1994
  (European Consortium for Mathematics in Industry ; Vol. 10)
  ISBN 978-3-322-96659-9   ISBN 978-3-322-96658-2 (eBook)
  DOI 10.1007/978-3-322-96658-2
NE: Engl, Heinz W. [Hrsg.] ; Inverse problems and optimal design
  in industry; European Consortium for Mathematics in Industry:
  European Consortium for ...

PREFACE

This volume contains thirteen papers and one extended abstract based on talks given at the symposium "Inverse Problems and Optimal Design in Industry" which took place from July 8 to 10, 1993, in Philadelphia. This symposium was jointly organized by ECMI an SIAM, with the cooperation of IMA (Minnesota), INRIA, and SIMAI. The organizing committee was co-chaired by the editors of this volume and included, in addition, V.Boffi (SIMAI, Italy) G. Chavent (INRIA, France), D. Colton (University of Delaware, USA) and A. Friedman (IMA, USA). Financial support from the U.S. Department of Energy is gratefully acknowledged.

The primary aim of this meeting, which is reflected in this volume, was to bring together mathematicians working in industry who treat inverse and optimal design problems in their practical work and mathematicians from academia who are active in mathematical research in these fields in order to strenghten the contacts between industry and academia. Thus, this volume contains (refereed) papers both on inverse and optimal design problems as they appear in European, American and Japanese industry, and on analytical and numerical techniques for solving such types of problems. The topics treated include multi-disciplinary design optimization in aerospace industry, inverse problems in steel industry, inverse and optimum design problems in optics and photographic science, inverse electromagnetic problems including impedance imaging, and inverse problems in the petroleum industry.

There were two styles of presentation of topics. One was discussion sessions. There were three of these on "Optimal Design" chaired by P.Neittaanmäki (University of Jyväskylä, Finland) and by J.Periaux (Dassault, France), on "Inverse Problems in Optics" chaired by J.A.Cox (Honeywell,USA) and by M.Maes (Philips,Netherlands), and on "Inverse Problems in Semi-Conductor Design" chaired by L.Borucki (Motorola,USA). There were also eighteen talks by G.R.Shubin (Boeing, USA), V.Shankar (Rockwell International, USA), C.Bischof (Argonne National Laboratory, USA), K.Yoda (Mitsubishi Electric, Japan), U.d'Elia (Alenia, Italy), A.Preuer (Voest Alpine Stahl, Austria), S.Halvorsen (Elkem Research, Norway), H.G.Stark (TecMath, Germany), L.Gurvits (Siemens Corporate Research, USA), L.A.Feldcamp (Ford Motor Co., USA), I.Hagiwara (Nissan Motor Co., Japan), D.S.Ross (Eastman Kodak, USA), M.Bertero (Universita di Genova, Italy), M.Vogelius (Rutgers Univsity, USA), D.Isaacson (Rensselaer Polytechnic Institute, USA), K.Baba (Mitsubishi Heavy Industries, Japan), C.Chardaire-Riviere (Institut Francais du Petrole, France), and R.E.Ewing (Texas A&M University, USA). Most of the speakers submitted papers to these Proceedings.

We would like to thank the members of the organizing committee for their efforts and, especially, the staff of SIAM for their efficient organization of the meeting.

Heinz W. Engl
Johannes Kepler Universität
Linz, AUSTRIA

Joyce McLaughlin
Rensselaer Polytechnic Institute
Troy, NY, USA

Table of Contents

INVERSE PROBLEMS IN PARTICLE SIZING AND CONFOCAL MICROSCOPY

M.Bertero and E.R.Pike

Abstract

Problems such as the determination of the sizes of microparticles by light scattering or the improvement of the images of a confocal scanning laser microscope require the solution of first-kind linear integral equations. In this paper we review our work intended to understand the limits of resolution which may be achieved in these problems. The results obtained are relevant not only for the correct interpretation of the output of commercial instruments but also for the improvement of their design.

1 Introduction

The scattering of a laser beam by a physical sample can be used for the determination of various properties of the sample such as diffusion coefficients, refractive index variations, fluid velocities and so on. This principle is the basis of a number of high precision optical instruments which are available commercially for research and production control. The main elements of these instruments are a laser and a photo-detector. Other specialized electronic circuits may be required for processing the detector output. Finally a computer is used for data interpretation. This last step, in most cases, implies the solution of a first-kind Fredholm integral equation.

As is well-known the mathematical problem of solving such an equation is ill-posed and this means in practice that there is strictly no unique solution which is compatible with the data, i.e. one can find many, quite different, solutions which fit the data provided by the instrument equally well. It may happen that the manufacturer does not emphasize this difficulty and, in order to have an instrument to sell, implements in the computer some *ad hoc* algorithm. In such a case the user is completely in the hands of the manufacturer and has no way to decide whether the solution provided by the computer is sound or not. In a few cases the manufacturer is more careful and allows the customer several algorithms to choose from.

In our opinion the important point is not so much to provide an algorithm for data inversion but first to provide methods for determining and quantifying the true information content of the data. In this paper we describe the results of a long-term collaboration intended to attack these problems using methods and concepts developed in the field of communication theory and radar signal processing. We also observe that a precise knowledge of the information content of a given experimental configuration permits not only a correct interpretation of the data but also the design of instrumentation to be carried out with the most efficient optical and electronic systems, with important advantages in cost and speed.

The first commercial light-scattering instruments based on the principles of the work described here have been recently developed and delivered (Real Time Granulometer series, Sematech, Sarl, Nice, France).

2 Particle sizing

The problem of particle sizing we consider is that of the estimation of the size of macromolecules or micro-particles, i.e. hydrosols or aerosols. In this field there are two important size ranges which determine the technique to be used for measurement.

The first is the *sub-micron range*, from small molecules ($\sim$ 50 Angström) up through macromolecular sizes to particles of a few microns in diameter. The second is the range of *larger sizes* which just overlaps the sub-micron region at its lower end and goes up to particles of the order of a millimetre in diameter at its maximum. The types of particles requiring measurement in these two ranges by particular industries or laboratories are too vast to enumerate but, for example, in the first range they encompass proteins, viruses, enzymes, colloids, micelles, latexes, inks and polymers while in the second range they encompass paints, cements, emulsions, fuel sprays, droplets and crystallites.

The division between the two ranges occurs at a particle radius roughly equal to the wavelength of visible light. In the larger size range, by definition, individual particles can be seen under the microscope and some instruments are based on the analysis of such images. If one has a mixture of particles of different sizes, one can count the number of particles for each size range and plot these numbers versus particle size. In this way the statistical distribution of particle sizes can be estimated.

A more statistically significant and more easily automated technique is based on *light scattering*. For the larger sizes this consists in measuring the intensity, as a function of angle, of monochromatic laser light scattered by all the particles in the laser beam (either in hydrosols or aerosols). This light is scattered into a cone in the forward direction and its intensity is measured in the focal plane of a high-aperture lens. By diffraction theory the light pattern is more closely concentrated near the axis for larger particles than for smaller ones.

For example, if we have identical particles of radius a and if a is larger than the wavelength λ of the laser light so that Fraunhofer approximation holds true, then the intensity of the scattered light $I(\vartheta)$, integrated over the azimuth angle, as a function of the scattering angle ϑ is proportional to[1]

$$g(\vartheta) \; = \; \frac{J_1^2(ka\vartheta)}{a\vartheta} \tag{2.1}$$

where

$$k \; = \; \frac{2\pi}{\lambda} \tag{2.2}$$

is the wave number. Then, if ϑ_1 is the first zero of $I(\vartheta)$ and if x_1 is the first zero of the Bessel function $J_1(x)$, by measuring ϑ_1 one can determine a through the formula

$$a \; = \; \frac{x_1}{k\vartheta_1} \; . \tag{2.3}$$

When ka is of the order of or smaller than unity, Mie scattering theory corrections must be taken into account.

In the case of two kinds of particles with radius a_1 and a_2 and probability f_1 and f_2 respectively, then the scattered intensity is proportional to

$$g(\vartheta) \; = \; f_1 \frac{J_1^2(ka_1\vartheta)}{a_1\vartheta} \; + \; f_2 \frac{J_1^2(ka_2\vartheta)}{a_2\vartheta} \; . \tag{2.4}$$

More generally, if we have a continuous distribution of particle radii, with probability density $f(a)$, the measured quantity $g(\vartheta)$ is given by

$$g(\vartheta) \; = \; \int_0^{+\infty} \frac{J_1^2(ka\vartheta)}{a\vartheta} \; f(a)da \tag{2.5}$$

and therefore the problem is the estimation of $f(a)$ from given values of $g(\vartheta)$.

In the sub-micron region the scattered light intensity varies little with scattering angle and, for this reason, in recent years the most widespread method in this range of sizes makes use of another physical phenomenon, namely, that of the Brownian motion of the particles in liquid suspension[12]. We should mention, however, that the small variation with scattering angle which remains in the case of light scattering, together with the concentration dependence of the scattered intensity, also provides, with modern lasers and detectors, a good method of determination of molecular weight known as the Zimm-plot technique.

The method based on the Brownian motion of particles is especially useful in the case of hydrosols but it has also been recently used in the case of aerosols[15],[11]. It works as follows. A coherent light beam from a laser is used to illuminate the particles and interference between the light scattered by each of the particles creates a resultant intensity in the focal plane (this time of a much lower aperture lens) which has the form of a random *speckle pattern*. This fluctuates on a time scale given by the speed of diffusion, which is faster the smaller the particles. Typically the patterns will fade and evolve to completely different ones in anything from some tens of microseconds to a few milliseconds. In this case the analysis is thus of a pattern varying in time rather than in space as in the larger-size range. The motion of the speckle pattern is nowadays always analysed by *Photon Correlation Spectroscopy* (PCS) techniques[12] which were developed in the laboratory of one of the authors (E.R.Pike) in the late sixties. PCS involves high-speed digital electronic processing (correlation) of single light quanta (photons) detected by a sensitive photomultiplier detector situated at a point in the speckle pattern. The time decay of the speckle pattern shows up as a decay of the digital correlation function in the delay-time variable.

More precisely, if $I(t)$ is the intensity of the speckle pattern as a function of time, and if $g^{(2)}(\tau)$ is the normalized autocorrelation function of $I(t)$

$$g^{(2)}(\tau) = \frac{E\{I(\tau)I(0)\}}{E\{I(\tau)\}} \tag{2.6}$$

where E denotes expectation, then the following relation for Gaussian amplitude fluctuations, due to Siegert, holds true

$$g^{(2)}(\tau) = 1 + |g^{(1)}(\tau)|^2 \tag{2.7}$$

where $g^{(1)}(\tau)$ is the light-field amplitude correlation function. In the case of spherical particles of radius a, one has

$$|g^{(1)}(\tau)| = e^{-\Gamma\tau} \tag{2.8}$$

where Γ, the half-width of the field-fluctuation power spectrum, is the *line-width* of the optical spectrum and is related to the diffusion coefficient of the particles, D, and to the momentum transfer of the scattered light, K, as follows

$$\Gamma \;=\; DK^2 \;.\tag{2.9}$$

According to the Stokes-Einstein relation, the diffusion coefficient D is given by

$$D \;=\; \frac{k_B T}{6\pi\eta a}\tag{2.10}$$

where k_B is the Boltzmann constant, T the absolute temperature, η the viscosity coefficient and a the hydrodynamic radius of the particles. If K, T and η are known, by data fitting one can determine Γ, i.e. the time constant of the exponential (2.8), and therefore a, eqs.(2.9)-(2.10).

Again, if we have a continuous distribution of particle radii, with probability density $f(\Gamma)$ of the corresponding line-widths, the correlation function $|g^{(1)}(\tau)|$ is given by

$$|g^{(1)}(\tau)| \;=\; \int\limits_{0}^{+\infty} e^{-\tau\Gamma} f(\Gamma)dt \;.\tag{2.11}$$

Therefore the problem of determining $f(\Gamma)$ from measured values of $|g^{(1)}(\tau)|$ is that of *Laplace transform inversion*.

3 Resolution analysis and singular system analysis of particle sizing

Both eq.(2.5) and eq.(2.11) are examples of integral equations of the following kind

$$g(x) \;=\; \int\limits_{0}^{+\infty} K(xy)f(y)dy\tag{3.1}$$

which are closely related to convolution integral equations. In order to introduce the main concepts used for an information theory approach to the solution of these integral equations, we first consider the case of convolution equations.

In the case of noisy data the problem is the estimation of a function f such that

$$g = K * f + h \tag{3.2}$$

where g, the data function, is known but h, the function describing noise or experimental errors, is unknown except for the fact that it must be *small* in some norm, i.e. $\|h\| \leq \varepsilon$. By Fourier transforming both sides of eq.(3.2) one gets

$$\hat{g}(\omega) = \hat{K}(\omega)\hat{f}(\omega) + \hat{h}(\omega) \ . \tag{3.3}$$

Then, if f is also bounded with respect to some norm, $\|f\| \leq E$, according to general results of regularization theory[16], for all the frequencies ω such that

$$|\hat{K}(\omega)| \ \leq \ \frac{\varepsilon}{E} \tag{3.4}$$

it is possible to estimate $\hat{f}(\omega)$ by the ratio $\hat{g}(\omega)/\hat{K}(\omega)$. Therefore, if $\omega_{\max}$ is the maximum value of the modulus of the frequencies satisfying condition (3.4) (we assume that $|\hat{K}(\omega)| \to 0$ monotonically when $|\omega| \to \infty$), an estimate of f, f_{est}, is given by

$$f_{\mathrm{est}}(x) = \frac{1}{2\pi} \int_{|\omega| \leq \omega_{\max}} \frac{\hat{g}(\omega)}{\hat{K}(\omega)} \ e^{ix\omega} \ dx \ . \tag{3.5}$$

This function is band-limited and can be represented by the Whittaker - Shannon sampling expansion[14], the distance between sampling points being given by

$$d = \frac{\pi}{\omega_{\max}} \ . \tag{3.6}$$

Therefore one can say that the data provide an approximation of the unknown function f with a resolution given by eq.(3.6). *This resolution depends on the signal-to-noise ratio* E/ε save for the very special kernel sinc(x) of classical information theory where the information content is essentially independent of noise.

A similar analysis can be applied to eq.(3.1) by using the Mellin transform. For a square-integrable function defined on $\mathbb{R}_+$, the Mellin transform and its inversion formula are given by[19]

$$\bar{f}(\xi) = \int_0^{+\infty} f(x) \ x^{-\frac{1}{2}+i\xi}dx \ , \quad f(x) = \frac{1}{2\pi} \int_{-\infty}^{+\infty} \bar{f}(\xi)x^{-\left(\frac{1}{2}+i\xi\right)}d\xi \ . \tag{3.7}$$

If we take into account that, in the analysis of experimental data, a noise term must be added to the r.h.s. of eq.(3.1), then by taking the Mellin transform of both sides of this equation we get

$$\tilde{g}(\xi) \;=\; \tilde{K}(\xi)\tilde{f}(-\xi) + \tilde{h}(\xi) \; . \tag{3.8}$$

Again, if $\xi_{\max}$ is the maximum value of the Mellin frequencies satisfying the condition

$$|\tilde{K}(\xi)| \;\leq\; \frac{\varepsilon}{E} \tag{3.9}$$

an estimate of f is given by

$$f_{\text{est}}(x) \;=\; \frac{1}{2\pi} \int\limits_{|\xi|\leq\xi_{\max}} \frac{\tilde{g}(-\xi)}{\tilde{K}(-\xi)} \; x^{-\left(\frac{1}{2}+i\xi\right)}d\xi \; . \tag{3.10}$$

This function can be represented by an extension of the Whittaker-Shannon expansion, which is known as an *exponential-sampling expansion*[17]. The sampling points form a geometric progression with ratio

$$\delta \;=\; \exp\left(\frac{\pi}{\xi_{\max}}\right) \; . \tag{3.11}$$

The meaning of δ is that *it is impossible to resolve two different kinds of particles if the ratio between their sizes is less than δ*. The parameter δ is called the *resolution ratio* and it quantifies the amount of information which can be extracted from the data.

For example, for $E/\varepsilon = 100$, in the case of PCS, i.e. eq.(2.11), one has $\delta = 2.44$, while in the case of diffraction scattering, i.e. eq.(2.5), one has $\delta = 1.72$. The dramatic difference in the performance of the two kinds of measurement is evident. In fact, it has long been reluctantly accepted by the PCS community that very few parameters of a particle-size distribution are available from the data, no matter what algorithm is invoked for the calculation. The previous simple analysis explains why this is so in a rigorous and quantitative manner. This result will be confirmed by the subsequent, more refined analysis given below.

An important fact, first discovered experimentally[17] and then explained theoretically[3], is that an improvement of the previous resolution ratio is possible if *a priori* knowledge of the support of the unknown function is used in the inversion algorithm.

Information about the support can be obtained

a) in the case of PCS directly from the data; the first and second derivative of the Laplace transform at the origin provide the first and second moment of the unknown solution;

b) in the case of light scattering, from a preliminary inversion of the data performed without any assumption on the support.

The investigation of the effects of the knowledge of the support has suggested the investigation of several interesting mathematical problems.

3.1 Finite Laplace transform

If we know that the support of $f(x)$ is interior to the interval $[a, b]$, then the problem to be solved is:

$$g(x) = \int_a^b e^{-xy} f(y) dy .$$
(3.12)

This defines a compact operator from $L^2(a, b)$ into $L^2(\mathbb{R}_+)$, which is called *finite Laplace transformation*, and it is easy to prove that its singular values depend on a, b through the ratio[3]

$$\gamma = \frac{b}{a} > 1 .$$
(3.13)

It is possible to develop for this transform a theory partly analogous to that of Slepian for the finite Fourier transformation. In particular the singular functions of the finite Laplace transformation are also eigenfunctions of known linear differential operators[4] and can be easily computed. The first five singular functions in $L^2(1, \gamma)$ for $\gamma = 5$, are shown in Fig.1.

The possibility of improving the resolution ratio is controlled by the following parameter: let γ be the dilation factor of the interval, eq.(3.13), and let δ be the resolution ratio (3.11), then the number of sampling points interior to $[a, b]$ is the so-called generalized *Shannon number*

$$S = \frac{\ln \gamma}{\ln \delta} = \frac{1}{\pi} \xi_{\max} \ln \gamma ;$$
(3.14)

if S is not too large, then resolution beyond the limit δ is possible, because a significant extrapolation of the Mellin transform of $f(y)$ out of the band $[-\xi_{\max}, \xi_{\max}]$ is feasible.

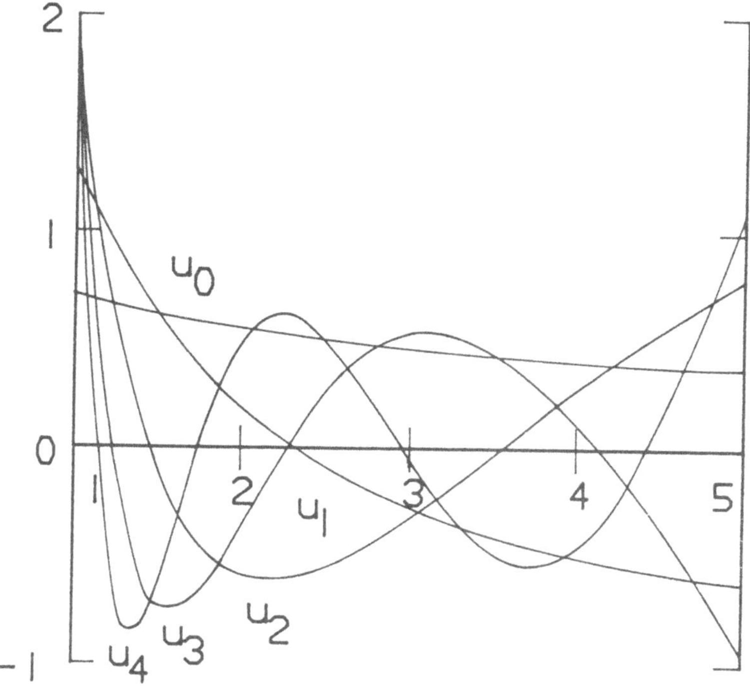

Fig.1 Plot of the first five singular functions $u_k(y)$ of the finite Laplace transformation in the case $\gamma = 5$.

We give an example of the estimation of this improvement using the singular values of the finite Laplace transformation. In the case $\gamma = 5$ and $\delta = 2.44$ (corresponding to $E/\epsilon = 100$) we have $S = 1.8$ and the first few singular values are

$$\sigma_0 = 0.8751 \quad , \quad \sigma_1 = 0.1935 \quad , \quad \sigma_2 = 0.0383 \quad , \quad \sigma_3 = 0.0074$$

so that we have 3 singular values greater than 10^{-2}. It can be shown that the possibility of estimating three components of $f(y)$ in the singular-function basis corresponds to the possibility of estimating three sampling points, instead of $S \sim 2$, interior to the support. The corresponding δ is 1.71 and this implies an improvement of 30% with respect to the inversion without knowledge of the support.

Unfortunately the inversions based on the use of these singular functions produce artifacts because they are large at the boundaries of the support (see Fig.1), where the solution is expected to be small.

3.2 An inversion in a weighted space

In order to overcome the difficulties due to the behaviour of the singular functions mentioned above, the problem of inverting the Laplace transformation has been considered by assuming that f belongs to a weighted L^2-space[5], i.e.

$$\|f\|^2 \;=\; \int\limits_{0}^{+\infty} \frac{|f(y)|^2}{P^2(y)}\; dy \;<\; \infty \; . \tag{3.15}$$

The *profile function* $P(y)$ contains the information about the support of $f(y)$, in particular it is small where $f(y)$ is known to be small. Under broad conditions on $P(y)$, the Laplace transformation is again a compact operator and its singular functions (in the solution space) contain a factor $P^2(y)$. Therefore inversion algorithms based on singular-function expansions provide solutions which are small precisely where they are assumed to be small. These reconstructions, of course, will depend on the choice of $P(y)$, but if this profile is well chosen this method provides satisfactory results with a resolution as that predicted by the method described in Sect.3.1.

3.3 Truncated exponential-sampling expansions

As explained above both the data and the estimate of the unknown function can be approximated by means of the exponential-sampling expansion

$$F(x) \;=\; \sum_{n=-\infty}^{+\infty} F(x_n) S_\Omega \left(\frac{x}{x_n}\right) \tag{3.16}$$

where

$$x_n \;=\; \exp\left(\frac{\pi n}{\Omega}\right) \; , \quad S_\Omega(x) \;=\; \frac{1}{\sqrt{x}}\; \frac{\sin(\Omega \ln x)}{\Omega \ln x} \; . \tag{3.17}$$

This can be obtained from the Whittaker-Shannon expansion for band-limited functions by means of a simple change of variables[17]. In the case of (3.10) one has $\Omega = \xi_{\max}$.

The idea is to discretize the integral equation by representing both the data function and the solution by means of this expansion, but with different values of the band-width Ω[8]. As concerns data one can take the band-width $\Omega = \xi_{\max}$ while for the solution one can take a value of $\Omega(> \xi_{\max})$ as estimated from the singular-value analysis described in 3.1. Therefore this value depends not only on ε/E but also on γ. Finally the expansions are truncated by taking only the sampling points interior respectively to the data and solution supports.

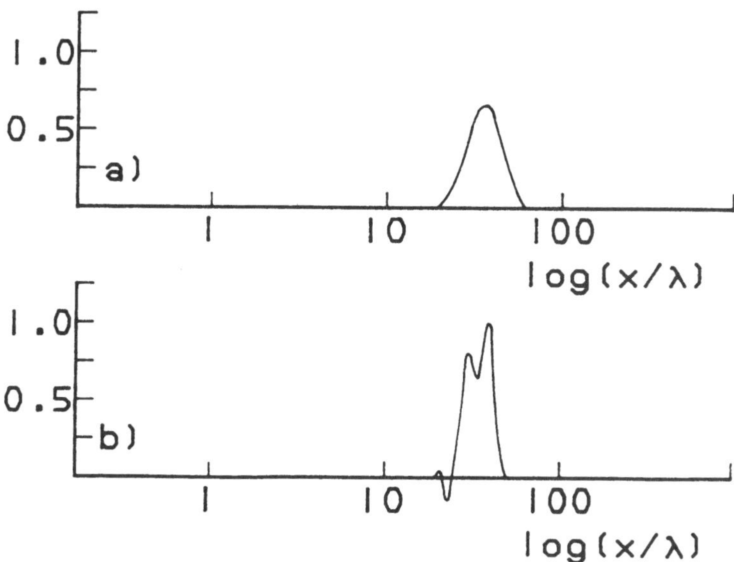

Fig.2 Inversion of light-scattering data corresponding to two delta functions at $x_1 = 30\lambda$ and $x_2 = 39\lambda(x_2/x_1 = 1.3; \lambda =$ wavelength of the incident radiation): a) inversion by means of the singular system corresponding to the support [1.25, 1.65]; b) inversion by means of the singular system of the support [20, 50] (see [8]).

The method has been applied both to the case of PCS (Laplace transform) and to the case of small-angle diffraction[8]. It can be easily implemented in order to realize so-called *zooming*: one first inverts the data using a broad support $[a_0, b_0]$ for the solution. If the restored solution is significantly different from zero only on a smaller interval $[a_1, b_1] \subset [a_0, b_0]$, then one can invert again the same data by restricting the sampling points to $[a_1, b_1]$ with a broader band-width. According to the theory one must find an improvement in resolution. An example is shown in Fig.2.

In conclusion the *resolution analysis* described above allows for

1) determination of an efficient and economic sampling scheme for the data (this implies an optimum design of the instrument);
2) definition of the resolution achievable using the relationship between resolution, signal-to-noise ratio and support of the unknown function;
3) choice of the inversion method based on the resolution which can be theoretically obtained.

4 Confocal microscopy

An important recent advance in microscopy has been the development of the confocal laser scanning microscope[9],[18] in which a laser beam is tightly focussed onto a specimen by a high-numerical-aperture *illumination* lens and viewed through a small pinhole in the image plane of a second such *imaging* lens. In the case of reflection microscopy, both illumination and imaging are effected with the same lens. Scanning is achieved by moving either the specimen or the laser beam (or both) in the lateral directions and by moving the specimen in the axial direction. The term *confocal* refers to the fact that the specimen is placed at the confocal point of the two lenses.

Commercial confocal microscopes are already produced. These are the first instruments which allow for 3D imaging of intact microscopic objects and therefore they are of great importance in biology and medicine where they can provide information about the structures of tissues, cells or cell nuclei. This possibility arises from the capability of the confocal arrangement to measure the density of fluorochromes in a small volume around the confocal point, with dimensions comparable with the wave length of the illuminating beam, rejecting contributions from other regions[10].

If we introduce optical coordinates v_x, v_y, u then the basic imaging equation in the case of 3D objects is given by[7]

$$g(\mathbf{v}) = \int W_2(|\mathbf{v} - \mathbf{v}'| \; ; \; u') \, W_1(|\mathbf{v}'| \; ; \; u') f(\mathbf{v}' \; ; u') \, d\mathbf{v}' \, du' \qquad (4.1)$$

where $W_1(|\mathbf{v}| \; ; \; u)$ and $W_2(|\mathbf{v}| \; ; \; u)$ are the intensity distributions in the focal region respectively of the illuminating lens and of the imaging lens. The function $g(\mathbf{v})$ is the image of the object $f(\mathbf{v}\ ;\ u)$ for a given scanning position; in practice it is the image of the values of $f(\mathbf{v};u)$ in a small volume around the point $\mathbf{v} = 0, u = 0$ and therefore $g(\mathbf{v})$ (which is a 2D image of a 3D object) is significantly different from zero only in a small area around $\mathbf{v} = 0$. In confocal microscopy the value of $g(\mathbf{0})$ (more precisely the integral of the values of $g(\mathbf{v})$ over a small disk around $\mathbf{v} = 0$, if one takes into account the finite size of the pinhole) is taken as an *image* of $f(\mathbf{0};0)$. A complete image of f is obtained by scanning, i.e. by translating the object $f(\mathbf{v};u)$, and one can easily see that the image obtained in such a way is given by

$$G(\mathbf{v};u) \ = \ \int T\left(|\mathbf{v} - \mathbf{v}'|;u'\right)f(\mathbf{v}';u')dv'du' \tag{4.2}$$

where

$$T(|\mathbf{v}|;u) \ = \ W_2(|\mathbf{v}|,u)W_1(|\mathbf{v}|;u) \ . \tag{4.3}$$

The improvement of resolution obtained by means of confocal microscopy is related to the fact that the band-width of $T(|\mathbf{v}|;u)$ is broader than the band-width of $W_2(|\mathbf{v}|;u)$ or $W_1(|\mathbf{v}|;u)$, i.e. the band-width of a conventional microscope. The transfer function, however, i.e. the Fourier transform of $T(|\mathbf{v}|;u)$ is not constant over the band but, unfortunately, it tends to zero rather rapidly from the center towards the boundary. For this reason a regularized deconvolution has been attempted[7] in order to improve the quality of the images. The result was that in the restored image some of the out-of-focus details of the object disappear and this provides an improvement of the axial resolution. Although in this method no extra information is used, in general the restored image is characterized by an apparent improvement in constrast and edge definition.

A method for obtaining a significant improvement of the resolution of a confocal microscope by recording extra information in the data was recently suggested by the authors[2]. This method is as follows:

1) detect the full image $g(\mathbf{v})$ (and not only $g(\mathbf{0})$) for a given scanning position;
2) solve the integral equation (4.1) to estimate $f(\mathbf{0};0)$ and take this estimate $f_{\text{est}}(\mathbf{0};0)$ as the new image of $f(\mathbf{0};0)$.

A complete image of $f(\mathbf{v};u)$, let us say $f_{\text{est}}(\mathbf{v};u)$, is obtained again by scanning and by repeating the previous procedures for each scanning position.

The previous method looks quite heavy from the computational point of view since it requires the solution of the integral equation (4.1). However we must observe that the integral operator does not depend on the scanning position and is compact (in L^2-spaces). Therefore the inversion algorithm (we have mainly considered truncated singular-function expansions) is the same for any scanning position. The singular system needs to be computed only once.

Numerical and analytical investigations of the singular systems of integral operators related to confocal microscopy have been accurately performed so that their main properties are now rather well known[13),6)].

We only sketch here the main feature of an inversion algorithm based on a truncated expansion since this has suggested an implementation of the algorithm by means of an optical processor.

If we denote by $\tilde{f}_K(0;0)$ the estimate of the solution of eq.(4.1) obtained by means of K singular functions we obtain

$$\tilde{f}_K(0;0) \; = \; \int \; M_K(\mathbf{v})g(\mathbf{v})dv \tag{4.4}$$

where

$$M_K(\mathbf{v}) \; = \; \sum_{k=1}^{K} \; \frac{1}{\sigma_k} \; U_k(0;0) \, V_k(\mathbf{v}) \; . \tag{4.5}$$

The σ_k are the singular values of the integral operator (4.1) and the $V_k(\mathbf{v})$ and $U_k(\mathbf{v}';u')$ are the associated singular functions respectively in the data and solution spaces.

Now the computation of $\tilde{f}_K(0;0)$ by means of formula (4.4) implies operations which can be performed by means of optical devices. In fact multiplication of $g(\mathbf{v})$ by $M_K(\mathbf{v})$ can be performed by placing a suitable optical mask (whose transparency is $M_K(\mathbf{v})$) in the image plane, so that, if $g(\mathbf{v})$ is the intensity at the entrance of the mask, $M_K(\mathbf{v})g(\mathbf{v})$ is the intensity at the exit of the mask. Integration is performed by means of a large detector. For incoherent microscopy the real situation is not so simple because one needs two masks, one corresponding to the positive values of $M_K(\mathbf{v})$ and the other to the negative values of $M_K(\mathbf{v})$[20)], but it is not necessary to discuss these details in this survey. A prototype of this optical processor has been already fabricated and tested at King's College, London and it has provided the expected improvement of resolution[21)].

In fact it can be shown that the system consisting of the confocal microscope and the optical processor described above provides an estimate $\tilde{f}_K(\mathbf{v};u)$

of $f(\mathbf{v}';u')$ which is given by

$$\tilde{f}_K(\mathbf{v};u) = \int T_K(\mathbf{v}-\mathbf{v}';u-u')f(\mathbf{v}';u')d\mathbf{v}'du' \qquad (4.6)$$

where

$$T_K(\mathbf{v};u) = \sum_{h=1}^{K} U_k(\mathbf{0};0)U_k(\mathbf{v};u) \qquad (4.7)$$

the singular functions U_k being the same as in eq.(4.5).

Eq.(4.7) must be compared with eq.(4.5). It turns out that the first term in eq.(4.7) has approximately the same behaviour as the function $T(|\mathbf{v}|;u)$ and therefore provides the same resolution. The other terms of eq.(4.7) are larger than the first one at higher frequencies and therefore are responsible of the improvement of resolution provided by the new microscope.

No commercial implementation of the super-resolving confocal microscope is yet available but we are actively developing a full 3D, high aperture prototype.

Aknowledgements

We are grateful for support for this programme from NATO, EEC and SERC as well as from italian MURST and INFM. We would also like to aknowledge many colleagues who have contributed from our own and other laboratories who may be identified by coauthorship in our list of references.

References

[1] Bayvel, L.P.; Jones, A.R.: Electromagnetic Scattering and its Applications. Applied Science Publishers (1981).

[2] Bertero, M.; Pike, E.R.: Resolution in diffraction limited imaging, a singular value analysis I. The case of coherent illumination. Optica Acta <u>29</u> (1982) 727-746.

[3] Bertero, M.; Boccacci, P.; Pike, E.R.: On the recovery and resolution of exponential relaxation rates from experimental data: a singular-value analysis of the Laplace transform inversion in the presence of noise. Proc. R. Soc. Lond. <u>A383</u> (1982) 15-29.

[4] Bertero, M.; Grünbaum, F. A.: Commuting differential operators for the finite Laplace transform. Inverse Problems <u>1</u> (1985) 181-192.

[5] Bertero, M.; Brianzi, P.; Pike, E.R.: On the recovery and resolution of exponential relaxation rates from experimental data: Laplace transform inversions in weighted spaces. Inverse Problems $\underline{1}$ (1985) 1-15.

[6] Bertero, M.; Boccacci, P.; Defrise, M.; De Mol, C.; Pike, E.R.: Super-resolution in confocal scanning microscopy II. The incoherent case. Inverse Problems $\underline{5}$ (1989) 441-461.

[7] Bertero, M.; Boccacci, P.; Brakenhoff, G.J.; Malfanti, F.; van der Voort, H.T.M.: Three-dimensional image restoration and super-resolution in fluorescence confocal microscopy. J. Microsc. $\underline{157}$ (1989) 3-20.

[8] Bertero, M.; Pike, E.R.: Exponential-sampling method for Laplace and other dilationally invariant transforms: I. Singular-system analysis. Inverse Problems $\underline{7}$ (1991) 1-20; II. Examples in photon correlation spectroscopy and Fraunhofer diffraction. Inverse Problems $\underline{7}$ (1991) 21-41.

[9] Brakenhoff, G.J.; Blom, P.; Barends, P: Confocal scanning light microscopy with high aperture immersion lenses. J. Microsc. $\underline{117}$ (1979) 219-232.

[10] Brakenhoff, G.J.; van der Voort, H.T.M.; van Sprosen, E.A.; Nanninga, N.: Three-dimensional imaging by confocal scanning fluorescence microscopy. Ann. N. Y. Acad. Sci. $\underline{483}$ (1986) 405-415.

[11] Chowdhury, D.P.; Sorensen, C.M.; Taylor, T.W.; Merklin, J.F.; Lester, T.W.: Application of photon correlation spectroscopy to flowing Brownian motion systems. Appl. Optics $\underline{23}$ (1984) 4149-4154.

[12] Cummins, H.Z.; Pike, E.R. (editors): Photon Correlation and Light Beating Spectroscopy. Plenum Press (1977).

[13] Gori, F.; Guattari, G.: Signal restoration for linear systems with weighted impulse. Singular value analysis for two cases of low-pass filtering. Inverse Problems $\underline{1}$ (1985) 67-85.

[14] Jerri, A.J.: The Shannon sampling theorem. Its various extensions and applications: A tutorial review. Proc. IEEE $\underline{65}$ (1977) 1565-1596.

[15] King, G.B.; Sorensen, C.M.; Lester, T.W.; Merklin, J.F.: Photon correlation spectroscopy used as a particle size diagnostic in sooting flames. Appl. Optics $\underline{20}$ (1982) 976-978.

[16] Miller, K.: Least squares methods for ill-posed problems with a prescribed bound. SIAM J. Math. Anal. $\underline{1}$ (1970) 52-74.

[17] Ostrowsky, N.; Sornette, D.; Parker, P.; Pike, E.R.: Exponential sampling method for light scattering polydispersity analysis. Optica Acta $\underline{28}$ (1981) 1059-1070.

[18] Sheppard, C.J.R.; Choudhury, A.: Image formation in the scanning microscope. Optica Acta $\underline{24}$ (1977) 1051-1073.

[19] Titchmarsh, E.C.: Introduction to the Theory of Fourier Integrals. Clarendon Press (1948).

[20] Walker, J.G.; Pike, E.R.; Davies, R.E.; Young, M.R., Brakenhoff, G.J.; Bertero, M.: Superresolving scanning optical microscopy using holographic optical processing. J. Opt. Soc. Am. $\underline{A10}$ (1993) 59-64.

[21] Young, M.R.; Jiang, S.H.; Davies, R.E.; Walker, J.G.; Pike, E.R.; Bertero, M.: Experimental confirmation of superresolution in coherent confocal scanning microscopy using optical masks. J. Microsc. $\underline{165}$ (1992) 131-138.

M.Bertero, Dipartimento di Fisica, Università di Genova, and I.N.F.N.,
Via Dodecaneso 33, I-16146 Genova, Italy

E.R.Pike, Department of Physics, King's College,
Strand, London WC2R 2LS, U.K.

Inverse and Optimal Design Problems for Imaging and Diffractive Optical Systems

J. Allen Cox
Honeywell Technology Center

1. Introduction

Historically, the general problem of optimal design in optics has been concerned primarily with traditional imaging optical systems found, for example, in cameras, telescopes, and microscopes. In recent years the area has been broadened both by the need for improved, more efficient nonimaging optical systems, such as illuminators and backlights, and by the infusion of a new technology to fabricate diffractive phase structures. The design problem for nonimaging optical systems offers it own set of special challenges for mathematicians to such a degree that for many years engineers had to rely on empirical and simple analytical approaches. More recently, however, progress has been made in more rigorous mathematical approaches for the nonimaging problem, and these are described by Maes[18] in another article in this Proceedings. This paper addresses the need for optimal design of imaging and diffractive optical systems from the engineering viewpoint and the author's view of fruitful ground for applied mathematicians to explore.

2. Conventional Imaging Optical Systems

An imaging optical system maps a scene in object space into its conjugate in image space within prescribed tolerances for image quality for every point in the scene, as illustrated generically in Figure 1. A conventional imaging system performs this operation using only refractive and reflective elements (ordinary lenses, prisms, and mirrors) and dielectric coatings. The problem of optimal design for this class of systems is very old, dating back at least two hundred years. Excellent descriptions of the problem can be found in many sources (Born and Wolf[4] is a well known "standard").

The most common approach used for design of conventional optics is raytracing and is based on the approximations of geometrical optics. This approach is derived from the rigorous theory embodied in Maxwell's equations for the electromagnetic field in the limit of vanishing wavelength and a scalar field. In this limit one derives the eikonal equation describing the phase propagation of the electromagnetic field,

$$\text{grad}S \cdot \text{grad}S = n(x)^2 \, ,$$

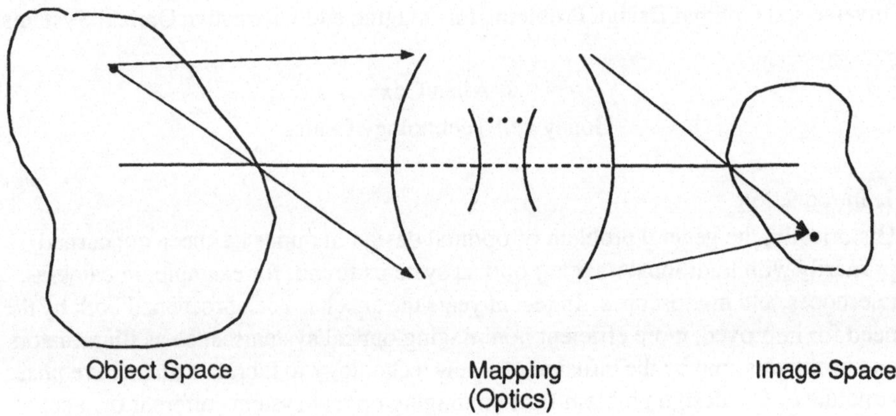

Figure 1. Transformation performed by an imaging optical system

where $S(x)$ represents the phase of the electromagnetic field and $n(x)$ is the index of refraction of the medium in which the wave propagates. The surface $S(x) = $ constant is the geometrical wavefront; an optical "ray" is just the local normal at one point on the geometrical wavefront. Equations for a ray's trajectory and path length can be derived from the eikonal equation. Snell's Law, representing the bending of a ray in crossing the interface separating two homogeneous regions of refractive index n_1 and n_2,

$$n_1 \sin\theta_1 = n_2 \sin\theta_2 \, ,$$

is a well known special case derived from the ray equations.

Further approximations exist within the framework of geometrical optics. The oldest and most common is the paraxial approximation, in which only small object heights and ray angles are considered such that $\sin\theta \cong \theta$. The computational simplicity of this approximation led to its widespread use prior to the advent of modern, high-speed computers. Today, the paraxial theory finds it greatest utility as a means of defining "ideal" imagery, since in this approximation aberrations arising from higher order terms are vanishingly small and every geometrical point in object space is mapped into a conjugate geometrical point in image space.

With the current availability of modern computers, virtually all approaches for optimal design of conventional imaging optical systems are based on raytracing using the full approximation of geometrical optics. In this case all rays emanating from any geometrical point in object space generally do not intersect at a conjugate point in image space, but rather form a distribution around a point. This situation closely corresponds to reality; precise observation of images formed by point sources readily show the light distributed over regions that can deviate significantly from the "ideal" point image. In the parlance of optical design, these distributions of the light, in representing deviations from the ideal point image, are called aberrations and can be grouped into broad categories (e.g., spherical aberration, coma, astigmatism, distortion, etc.). Image quality is determined by the type and the magnitude of the aberrations over the field-of-view of the system. The primary objective in the optimal design of imaging optics is to find the minimal set of optical elements (lenses, mirrors, etc) which provides the required image quality over a specified field (i.e., over some specified domain in either object space or image space).

Typically, design requirements are specified in the form of data such as angular field-of-view, spectral waveband, image size, focal length, aperture diameter, size constraints, and cost constraints. If one limits the design to only lenses and mirrors having planar or spherical surfaces, then the degrees of freedom in the design process are limited to: radius of curvature of each surface, refractive index of each lens element, dispersion in the refractive indices ($dn/d\lambda$), spacing of the elements, and placement of "stops" (apertures). Permitting the use of aspheric surfaces can significantly increase the degrees of freedom, but also the cost.

The design process usually starts with the selection of a merit function to measure image quality. Typically, the merit function represents some measure of the spread in the light distribution of a point source image. An initial optical configuration is selected from experience of the designer. A finite number of points spanning the field in object space and a finite of wavelengths covering the spectral waveband are selected, and, if appropriate, weighting factors for each are assigned. The distribution in image space conjugate to each point in object space is calculated by tracing a bundle of rays through the system. The problem of optimal design now consists of finding those regions in parameter space spanning the degrees of freedom in such a way that the merit function at all selected field points lies within required tolerance limits. It is at this point that the problem of inverse and optimal design for optics shares many similarities with other engineering disciplines and that potential opportunities exist for applied mathematicians. We return to this point in the last section.

3. Diffractive and Hybrid Optical Systems

Technology developments over the past 20 years in high precision micromachining techniques - such as laser-based and direct-write electron-beam lithography, reactive-ion-etch processing, x-ray lithography and LIGA processing, and single-point diamond machining - have permitted the creation of gratings and other diffractive phase structures with small periods and high aspect ratios. The power of these devices lies in the ability to implement rather general transformations on an incident wavefront. Figure 2 illustrates the various activities and processes involved in diffractive optics as a whole as practiced at Honeywell. Generally, three major activities are involved: design and performance analysis, master fabrication, and replication. The generality of the optical transformations that can be implemented with a diffractive element dramatically increases the degrees of freedom in the design process, leading to simpler designs with fewer elements. The ability to replicate the diffractive element with high fidelity and image quality keeps the cost per element small for large quantities. A hybrid optical system is one having both conventional and diffractive elements. A good overview of the whole technology can be found in the papers in Lee[17].

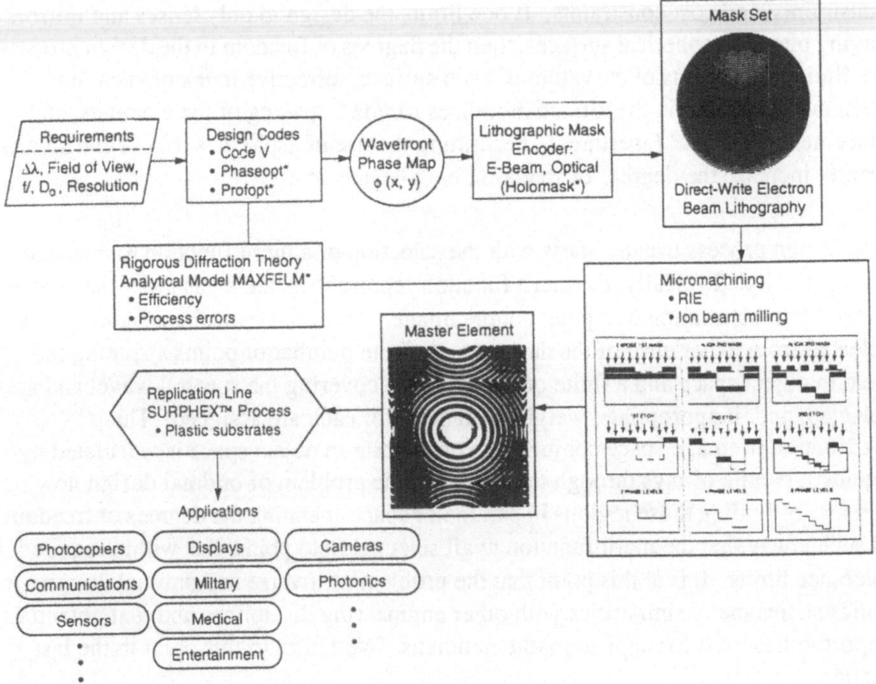

Figure 2. Overview of diffractive optics technology.

The practical application of diffractive optics technology for both imaging and nonimaging devices has driven the need for mathematical models and numerical codes both to provide rigorous solutions of the full electromagnetic vector-field equations for complicated grating structures, thus predicting performance given the structure, and to carry out optimal design of new structures. The direct problem and the inverse problem are both areas of active research in optical engineering community, although only infrequently does one group tend to do work in both areas. For the direct problem, several approaches have been taken. Among the most well known are the differential and integral methods described by Petit, Neviere et al.[19] and the differential, coupled-waves method of Gaylord and Moharam[14].

For the inverse problem, by far the greatest activity has been in global optimization techniques for raytracing and phase reconstruction techniques. Much of the raytracing optimization work is done by vendors of commercial optical design codes. Phase reconstruction is used primarily to design phase-only diffractive devices to produce a desired far-field intensity distribution. The Gerchberg-Saxton algorithm[15] yields reasonably high efficiency and has been most widely used for this purpose, although Farn[13] recently has demonstrated a new dual-loop algorithm that yields both good efficiency and intensity shaping over orders. These phase reconstruction methods are valid within the domain of Fourier optics; attempts to extend the validity further, either into the near-field or to "fast" optics(elements with features small compared to the optical wavelength), mostly have been confined, at least in the engineering and optics community, to techniques based on simulated annealing[16] and, very recently, on relaxed optimization[9,11].

Deeper understanding of the need for new directions in optimal design methods for diffractive optics can be found by considering the role of process errors on performance. As an example, one common method of micromachining a surface-relief phase profile into a substrate follows the "binary optics" strategy developed at MIT Lincoln Laboratory[20]. Here, a staircase approximation to the ideal continuous phase profile is fabricated in N mask-align-and-etch steps yielding 2^N phase levels. This procedure is indicated conceptually in the lower right box in Figure 2. Typically, performance measures, such as diffraction efficiency, of such binary optical elements is calculated only for geometrically perfect profiles without considering the unavoidable effects of processing errors. Such predictions thus represent an upper limit on performance expected for all elements.

The effects of processing errors on performance of these kind of diffractive optics (binary optics) have been investigated systematically by Cox et al.[7,8] at Honeywell. Processing errors can be classified into the following four major categories:
- etch depth errors
- linewidth errors (resulting from photoresist development)
- mask alignment errors

- Fresnel zone definition errors

A Fresnel zone is that region in the area of a diffractive element where the phase is bounded within one specific cycle. On the basis of extensive experimental and theoretical study[7,8], we have found that image quality is remarkably insensitive to these four errors, assuming magnitudes routinely achievable with existing technology. Diffraction efficiency in binary optics, however, is quite sensitive to errors resulting both from mask alignment and from photoresist linewidth at a mask alignment interface.

Figure 3 illustrates the effect of mask alignment error on the profile structure and shows how diffraction efficiency varies as a function of Fresnel zone width for four different values of mask alignment error. Analyses of this nature have led to the conclusion that with the binary optic fabrication procedure the minimum Fresnel zone width must exceed the feature error by a factor of one hundred. In other words, if the diffractive element has a minimum Fresnel zone width of 10 μm, then the mask alignment error must be less than 0.1 μm. Errors of this magnitude exceed the capability of manual mask aligners and can be attained only with interferometric methods implemented, for example, in direct-write electron beam machines.

4. <u>Role</u> for <u>Applied</u> <u>Mathematicians</u>

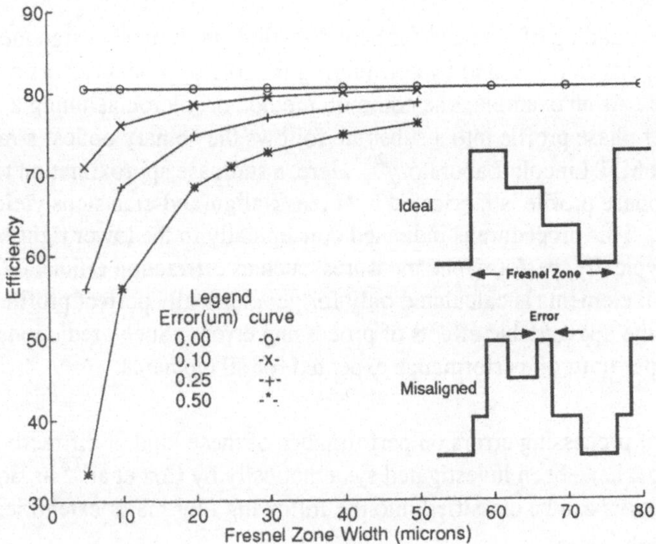

Figure 3. Effect of mask alignment error in a four level binary optical element.

Having reviewed in layman's terms (both from the mathematician's and the engineer's viewpoints) the situation in optimal design for imaging optical systems using conventional and diffractive optics, I would now like to suggest possible areas of collaboration between applied mathematicians and optical engineers. I shall then conclude with one specific example of a successful joint effort between industry and applied mathematicians.

In the area of optimal design for conventional optical imaging systems, there clearly is much common with similar problems for other engineering disciplines, and much of the activity in this area by applied mathematicians is undoubtedly relevant to optics. There are currently several optical raytracing computer codes available commercially, and the competition among the businesses supporting the different codes is keen. It is this very competition that drives the need to improve the raytracing codes. Two well advertised areas of improvement are global optimization techniques to avoid local minima in the merit function and efficient search algorithms to speed the convergence. Because of the competitive nature of this business, however, much of this work is supported internally by the private companies and is not reported publicly in the scientific literature. So, in this case there is a significant role for applied mathematicians, but there is also a significant level of activity.

In the case of optimal design for diffractive optics, the situation is somewhat different. In reviewing the optics literature since 1970 one is struck by the relatively small number of articles by applied mathematicians in this area. For the direct problem, the work of Maystre[19] and Cadilhac[6] are noteworthy exceptions, and recently Abboud and Nédélec[1] have described an variational approach similar to Dobson's[11]. The scarcity of direct involvement by applied mathematicians is remarkable, given the inherent nature of the problem, the complex issues encountered, and the number of open questions. The issues of existence and uniqueness of solutions have been raised already by Petit[19] and Cadilhac[6], and in the optics and engineering literature it is difficult to find any rigorous treatment of convergence. For the inverse problem, the situation is similar to that described above for raytracing. There is a large and active group of applied mathematicians working in this area, but the specific applications generally lie outside optics. There have been a number of developments in this area which can be applied profitably to optics.

As one attempt to draw applied mathematicians into this area, Honeywell and the National Science Foundation established an industrial postdoctoral position in diffractive optics at the Institute for Mathematics and Its Applications (IMA) at the

University of Minnesota in 1990. The collaboration between industry and applied mathematicians has proved to be fruitful and rewarding. Three new approaches for solving the direct problem have resulted the Honeywell/IMA association: an integral method[12], an elegant analytic continuation method[5], and a versatile, robust variational method[11]. Existence and uniqueness theorems have been proven[12,3], and convergence formulas have been derived for all three methods[12,5,2]. In the area of the inverse problem and optimal design, a nonlinear minimization approach using Newton's method has been successfully developed for phase reconstruction[10]. In addition, the more recent techniques of relaxed optimization have been applied to predict entirely new grating structures having antireflective or beam generation properties with the wave propagation described by the Helmholtz equation[9,11].

Mathematical modeling represents an important part of Honeywell's overall approach to fabricate master diffractive elements for subsequent replication. As part of a comprehensive development effort in diffractive optics technology, Honeywell established in 1990 an Industrial Postdoctoral Program in Diffractive Optics Modeling at the Institute for Mathematics and Its Applications (IMA) at the University of Minnesota. The objective of this program is to develop mathematical models for designing (the inverse problem) and analyzing (the direct problem) general periodic surface-relief structures such as diffraction gratings and to implement the models in computer codes using numerical algorithms with good convergence.

Three new computer models have resulted from the Honeywell/IMA postdoctoral program. The roles of these codes in Honeywell's overall diffractive optics technology thrust are indicated in Figure 2. Briefly summarized, these are:

1. MAXFELM
   - applicable to the direct problem
   - provides solution to full vector set of Maxwell's equations
   - biperiodic boundary with completely general dielectric properties
   - finite element implementation of variational method
   - iterative solution of discretized equations
2. PHASEOPT
   - applicable to the inverse problem
   - scalar field in the Fraunhofer approximation
   - biperiodic gratings with arbitrary intensity distribution in propagating orders
   - nonlinear least-squares method with conjugate-residual inner loop

3. PROFOPT
   - applicable to the inverse problem
   - scalar field supported by the full Helmholtz equation
   - singly periodic gratings

Typically, the two codes, PHASEOPT and PROFOPT, are applied to problems of optimal design to yield diffractive profiles which best meet a set of specified requirements. The structure must then be modeled with the MAXFELM code as a direct problem to examine its sensitivity to processing errors and to assess it degree of manufacturability. Recently, we have started an effort to extend the PROFOPT code to give diffractive structures which not only meet specified performance requirements but also comply with known manufacturing constraints. Our first applications have been directed both at angle-optimized antireflective "moth-eye" structures and at an ideal array generator. Especially for the latter example, where the grating period is somewhat larger, we have found promising results.

References

[1] Abboud, T., and Nédélec, J., "Electromagnetic waves in an inhomogeneous medium," J. Math. Anal. Appl. Vol. 164, pp. 40-58 (1992).

[2] Bao, G., "Finite elements approximation of time harmonic waves in periodic structures," IMA Preprint Series # 1156, submitted.

[3] Bao, G. and Dobson, D.C., "Diffractive optics in nonlinear media with periodic structure," IMA Preprint Series #1124, submitted.

[4] Born, M. and Wolf, E., Principles of Optics 6th Ed.(Permagon Press, NewYork, 1980).

[5] Bruno, O.P. and Reitich, F., "Numerical solution of diffraction problems: a method of variation of boundaries," J. Opt. Soc. Amer. A, to appear.

[6] Cadilhac, M., "Rigorous vector theories of diffraction gratings," in Progress in Optics Vol 21, E. Wolf, ed., (North-Holland, New York, 1984)

[7] Cox, J.A. et al., "Diffraction efficiency of binary optical elements," Proceedings SPIE Vol. 1211, pp. 116-24 (1990).

[8] Cox, J.A., Fritz, B.S., and Werner, T., "Process error limitations on binary optics performnce," Proceedings SPIE Vol. 1555, pp. 80-8, (1991).

[9] Dobson, D.C., "Optimal design of periodic antireflective structures for the Helmholtz equation," Euro. J. Appl. Math., Vol. 4, (1993) in press.

[10] Dobson, D.C., "Phase reconstruction via nonlinear least-squares," Inverse Prob-

lems Vol. 8, pp. 541-57 (1992).

[11] Dobson, D.C. and Cox, J.A., "Mathematical modeling for diffractive optics," in Diffractive and Miniaturized Optics, SPIE Critical Reviews Vol CR47, S.H. Lee, ed. (1993).

[12] Dobson, D. and Friedman, A., "The time-harmonic Maxwell equations in a doubly periodic structure," J. Math. Anal. Appl. Vol. 166, pp. 507-28 (1992).

[13] Farn, M.W., "New iterative algorithm for the design of phase-only gratings," Computer and Optically Generated Holographic Optics, I.N. Cindrich and S. Lee, eds., Proceedings SPIE, Vol. 1555, pp. 34-42 (1991).

[14] Gaylord, T.K. and Moharam M.G., "Analysis and Applications of Optical Diffraction by Gratings," IEEE Proceedings Vol. 73 No. 5, pp. 894-937 (1985).

[15] Gerchberg, R.W. and Saxton, W.O., "A Practical Algorithm for the Determination of Phase from Image and Diffraction Plane Pictures." Optik Vol. 35 No. 2, , pp. 237-46, (1972).

[16] Kirkpatrick, S., Gelatt, C.D. Jr.,Vecchi, M.P., "Optimization by Simulated Annealing," Science Vol. 220, pp. 671-80 (1983).

[17] Lee, S.H., ed., Diffractive and Miniaturized Optics, SPIE Critical Reviews, Vol. CR49 (SPIE, Bellingham WA, 1994).

[18] Maes, M., this Proceedings.

[19] Petit, R., ed., Electromagnetic Theory of Gratings (Springer-Verlag, Berlin, 1980).

[20] Veldkamp, W.B. and Swanson, G.J., "Developments in fabrication of binary optical elements," Proceedings SPIE Vol. 437, pp. 54-9 (1983).

Honeywell Technology Center
3660 Technology Drive
Minneapolis, MN USA 55518

INVERSE PROBLEMS IN ELECTROMAGNETISM: ANTENNA'S APPLICATIONS

Ugo F. D'Elia
Alenia
Rome, Italy

Abstract

Numerous examples of inverse problems can arise from the applications of Electromagnetic Theory. The purpose of this paper is to give an overview, from a mathematical point of view, of the actual problems in designing, developing and testing an "Antenna System". We are concerned with both "Inverse" problems: "Synthesis" (in design) and "Identification" (in measurements). The paper starts with a brief description of the basic mathematical modelling, and related numerical methods, of the Direct Problem: Antenna Analysis, then traditional Synthesis techniques are reviewed followed by a new generalized method. Last topic deal with inverse problems that comes from modern testing and measurement techniques. The difference between modern and old fashion design approach is also emphasised.

1. Introduction

An antenna is the system component that is designed to radiate and/or receive Electromagnetic (EM) waves. It can have a simple geometry:

wire antenna of different shapes,

aperture antenna like an horn or a truncated waveguide,

or can assume a more complex configuration:

reflector antenna made of a primary radiator (or feed) and one or more reflecting surfaces to collimate the energy, *array antenna* that is an aggregate of radiating elements regularly or irregularly distributed on a surface, usually planar. Special or high sophisticated array applications, called "*conformal*" require or use curved surfaces.

While in the past antenna design may have been considered a secondary issue in overall system design, today it plays a critical role. In fact, a good design of antenna can relax system requirements and improve overall performance. In the first half of this century, and probably till all the sixties, antenna technology may have been considered almost a "cut and try" operation. A good project needed the construction of several physical models often scaled in frequency. This full experimental method took too much time and cost giving often approximate results. Nowadays the analysis and design methods are such that antenna system performance can be predicted with remarkable accuracy making extensive use of computational methods and Computer Aid. In fact, many antenna designs proceed directly from the initial design stage to the prototype without intermediate testing. As a consequence, in the modern mathematical based approach, illustrated in fig. 1.1, inverse and direct sophisticated algorithms together with high reliable measurement technique are required to achieve an "Optimal Design".

In the complete theoretical and numerical first phase of the antenna design, "direct" and "inverse" scattering problems must be solved. In addition "optimization" technique are necessary for the dimensioning of many crucial system parameters. In the second experimental and numerical phase of the antenna project, "inverse" algorithms are needed to verify the performances and identify the primary antenna functions. In this phase, the antenna configuration is considered unknown and is to be "identified" from the measured scattering fields.

Before describing the synthesis methods, the following paragraph illustrates the direct problem and reports the mathematical models needed for the following synthesis.

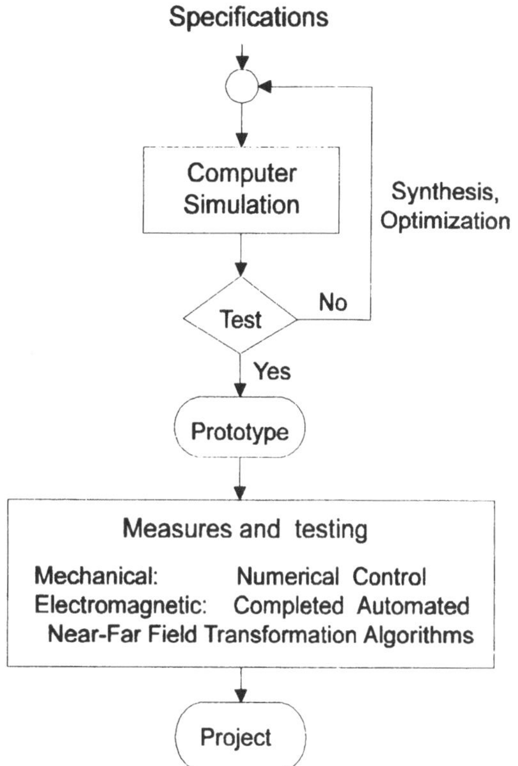

Fig. 1.1: Modern Design Approach: The Mathematical model

2. The direct problem: Antenna Analysis

The direct problem for an antenna system is an Electromagnetic boundary value problem.   It can be stated as follows:

given: the sources of excitation, their geometric distribution and the Electromagnetic characteristics of materials involved,

determine: the Electric and Magnetic Fields radiated.

The field radiated must satisfy Maxwell's equations, which, for a lossless medium ($\sigma=0$) and time-harmonic fields (assuming an $e^{j\omega t}$ time convention), can be written as:

$$\nabla \times \mathbf{E} = -\mathbf{M}_i - j\omega\mu\mathbf{H} \qquad \nabla \cdot (\varepsilon\mathbf{E}) = q_{ve}$$
$$\nabla \times \mathbf{H} = +\mathbf{J}_i + j\omega\varepsilon\mathbf{E} \qquad \nabla \cdot (\mu\mathbf{H}) = q_{vm} \tag{2.1}$$

Both electric ($\mathbf{J}_i$) and magnetic ($\mathbf{M}_i$) current densities, and electric (q_{ve}) and magnetic (q_{vm}) charge density represent the sources of excitation. Although magnetic sources are not physical, they are often introduced as electrical equivalents to facilitate solutions of physical boundary-value problems.   In addition to Maxwell's equations, the fields must satisfy the boundary conditions on he discontinuities:

$$\hat{n} \times \mathbf{E}_d = \mathbf{M}_s \quad \hat{n} \times \mathbf{H}_d = \mathbf{J}_s \quad \hat{n} \cdot (\varepsilon\mathbf{E}_d) = q_{se} \quad \hat{n} \cdot (\mu\mathbf{H}_d) = q_{sm} \tag{2.2}$$

and satisfy the "*radiation condition*" which requires, in an infinite homogeneous medium, that the waves travel outwardly from the source and vanish at infinity.

A common procedure to solve a radiation problem is based on the use of auxiliary functions: the "vector and/or scalar potentials". Common electric and magnetic vector potentials are $\mathbf{A}$ and $\mathbf{F}$, and scalar potential are ϕ_e and ϕ_m.   The definitions and their relations with $\mathbf{E}$ and $\mathbf{H}$ fields are:

$$\mathbf{H}_A = (1/\mu)\nabla \times \mathbf{A} \qquad \mathbf{E}_F = -(1/\varepsilon)\nabla \times \mathbf{F} \tag{2.3}$$

$$\mathbf{E}_A = -j\omega\,\mathbf{A} - \frac{j}{\omega\mu\varepsilon}\nabla(\nabla \cdot \mathbf{A}) \qquad \mathbf{H}_F = -j\omega\,\mathbf{F} - \frac{j}{\omega\mu\varepsilon}\nabla(\nabla \cdot \mathbf{F}) \tag{2.4}$$

$$\nabla \cdot \mathbf{A} = -j\omega\varepsilon\mu\,\phi_e \qquad \nabla \cdot \mathbf{F} = -j\omega\varepsilon\mu\,\phi_m \tag{2.5}$$

$$\nabla\phi_e = -\mathbf{E}_A - j\omega\mathbf{A} \qquad \nabla\phi_m = -\mathbf{H}_F - j\omega\mathbf{F} \tag{2.6}$$

Using the previous relations, in a source free region, we can obtain, for $\mathbf{A}$ and $\mathbf{F}$, the following vector potential equations:

41

$$\nabla^2 \mathbf{A} + \beta^2 \mathbf{A} = -\mu \, \mathbf{J}_i \qquad \nabla^2 \mathbf{F} + \beta^2 \mathbf{F} = -\varepsilon \, \mathbf{M}_i \qquad (2.7)$$

where $\beta^2 = \omega^2 \mu \varepsilon$. To get the previous simple forms is fundamental the use of the expressions (2.5) known as Lorentz conditions. The simplest solutions to the above vector potential equations can be written as an integral of the infinitesimal current elements:

$$\mathbf{A} = \frac{\mu}{4\pi} \iiint_{I'} \mathbf{J} \, \frac{e^{-j\beta R}}{R} dv \qquad \mathbf{F} = \frac{\varepsilon}{4\pi} \iiint_{I'} \mathbf{M} \, \frac{e^{-j\beta R}}{R} dv \qquad (2.8)$$

The steps of the general solving procedure are:
- Specify the electric and magnetic current density source $\mathbf{J}$ and $\mathbf{M}$,
- Find the vector potentials $\mathbf{A}$ and $\mathbf{F}$ using the integrals (2.8),
- Find $\mathbf{E_A}$, $\mathbf{H_A}$, $\mathbf{E_F}$, and $\mathbf{H_F}$ using the (2.3) and (2.4),
- Finally the total fields are given by:

$$\mathbf{E} = \mathbf{E}_A + \mathbf{E}_F = -j\omega \, \mathbf{A} - \frac{j}{\omega\mu\varepsilon} \nabla(\nabla \cdot \mathbf{A}) - \frac{1}{\varepsilon} \nabla \times \mathbf{F} \qquad (2.9)$$

$$\mathbf{H} = \mathbf{H}_A + \mathbf{H}_F = -j\omega \, \mathbf{F} - \frac{j}{\omega\mu\varepsilon} \nabla(\nabla \cdot \mathbf{F}) + \frac{1}{\mu} \nabla \times \mathbf{A} \qquad (2.10)$$

Previous procedure is valid in any region of space. If we are only interested in the "far fields" radiated by an antenna of finite dimensions, the electric and magnetic fields have only θ and ϕ component and can be approximated by the following relations:

$$\begin{aligned}
\mathbf{E}_A &\cong -j\omega \, \mathbf{A} & \mathbf{H}_A &\cong \frac{\hat{\mathbf{r}}}{\eta} \times \mathbf{E}_A = -j\frac{\omega}{\eta} \hat{\mathbf{r}} \times \mathbf{A} \\
\mathbf{H}_F &\cong -j\omega \, \mathbf{F} & \mathbf{E}_F &\cong -\eta \, \hat{\mathbf{r}} \times \mathbf{H}_F = -j\omega \eta \, \hat{\mathbf{r}} \times \mathbf{F}
\end{aligned} \qquad (2.11)$$

where $\eta = \sqrt{\mu/\varepsilon}$ is the intrinsic impedance of the medium.

A very simple application of the above procedure can be utilized in finding the field radiated by an "infinitesimal" dipole of length $l << \lambda$. The geometry of the problem is illustrated in fig 2.1.

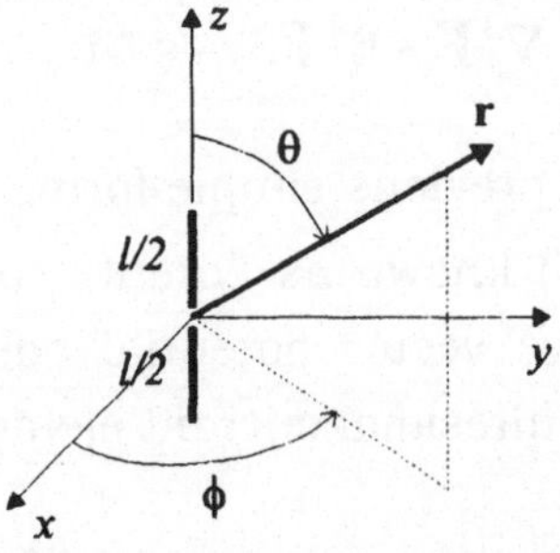

fig. 2.1

The current is assumed to be constant $\mathbf{I}(z) = \hat{\mathbf{z}}\, I_0$.   Following the procedures established, the vector potential can be written as:

$$\mathbf{A}(\dot{x},y,z) = \frac{\mu}{4\pi} \int I_0 \frac{e^{-jkR}}{R}\, d\mathbf{l} = \hat{\mathbf{z}}\, \frac{\mu I_0 l}{4\pi r}\, e^{-j\,kr}$$

With the relation (2.9) we can obtain the electric field components, that are valid everywhere, except on the source itself:

$$E_r = \eta \frac{I_0\, l\, cos\theta}{2\pi r^2}\left[1+\frac{1}{jkr}\right]e^{-jkr}, \quad E_\theta = j\,\eta \frac{kI_0\, l\, sin\theta}{4\pi r}\left[1+\frac{1}{jkr}-\frac{1}{(kr)^2}\right]e^{-jkr}$$

and $E_\varphi = 0$.   In the case of far field $(k\,r >> 1)$ we have:

$$E_\theta \cong j\,\eta \frac{kI_0\, l\, sin\theta}{4\pi r}\, e^{-jkr} \qquad (2.12)$$

Another fundamental technique used in electromagnetic modellization is the "Field Equivalence"; it is a principle, introduced by Schelkunoff [21], by which actual sources are replaced by "equivalent" electric or magnetic currents that, externally to a closed surface enclosing the actual antenna, produce the same fields as those radiated by the actual sources.   This principle leads to the wide used modellization of an antenna as a radiating "Aperture".

Previous analysis procedure is quite general and directly applicable only to simple configurations. For the calculus of the electromagnetic scattering of more complex structure, a suitable "modellization" by means of "canonical configurations" is necessary. To this end various theories and related algorithms give the numerical solution. In general, these methods are divided in two classes.

The asymptotic high frequency techniques (HFT) use an high degree of approximation and are based on the concept that the high frequency scattering is a "local phenomena".

The most common are the Geometrical Optics (GO), the Uniform Theory of Diffraction (UTD), the Physical Optics (PO), the Physical Theory of Diffraction (PTD) and the Method of Equivalent Current (MEC). All their numerical implementation make extensive use of sophisticated ray tracing algorithms. Between the first two (GO + UTD), that can be classified as ray optical approach, and the seconds (PO + PTD), that use a current integration approach, there is a method called Aperture Field Method (AFM). The last one, for his generality and simplicity is the most widely used as a model for traditional synthesis methods.

Another class of EM methods are based on the solution of Integral or Partial Differential Equations. The best known formulation are the "E-Field Integral Equations" (EFIE), the "H-Field Integral Equations" (HFIE) and the "Mixed Field Integral Equations" which is a combination of both EFIE and HFIE. The numerical procedures for this type of methods are the "Method of Moment" (MoM), the "Finite Element Method (FEM), Boundary Element Method (BEM), the Finite Difference method (FD) and mixed techniques. All these procedure discretize the problem and transform the continuous equation in a system of linear equation. Even if they are applicable, in principle, at any frequency, from a computational point of view, they can be used only in a low frequency range.

The method to adopt for predicting the radiation pattern depends on the type of the antenna. In general, low frequency techniques are used for small antennas, in term of wavelength, often used as single elements in Array antennas. The high frequency technique, instead, are useful for reflector antennas and general scattering problems.

3. The inverse problem: Antenna Synthesis

So far the attention has to been focused on Antenna Analysis that is the determination of the radiation pattern of a given antenna structure. The inverse of the analysis problem is usually referred to as "Synthesis". It can be expressed as the problem *to determine the geometrical and electromagnetic structure of the antenna so that it radiates in a prescribed manner and fulfils the required specifications.* Such definition is quite indefinite but it is inherent to its nature of an inverse problem and corresponds to the actual situation. Instead a direct problem computes the "effect" of a given cause and therefore is always well specified. This difference in the completeness of the definition is reflected to the computational methods. While the Antenna Analysis has many sophisticated numerical methods available in predicting the system behaviour, Synthesis tools are often fewer and consider only simplified formulations that face partial aspects of the antenna design. The following paragraph will give the basic ideas of the traditional synthesis methods. They can be of two types, the first one tries to fulfil the pattern specifications by varying the excitation of sources distributed on a fixed geometry, the second tries to fulfil the requirement modifying the geometry of the antenna with fixed primary excitation. In a second section a general unified framework for synthesis will be settled.

3.1 Synthesis: traditional methods

Most classical methods for the synthesis use the modellization of an antenna as a distribution of currents on an "Aperture". The currents, electric or magnetic, can be continuous or discretized in a linear or planar domain and are obtained, depending on the antenna and source configuration, usually by "equivalent principles". In the discretized case the antenna usually is an "Array" of simple radiating elements distributed on a regular lattice. As an example of continuous distribution the far field pattern of a finite length dipole ("linear aperture") is considered.

$$E_0 = j\eta \underbrace{\frac{k\,e^{-jkr}}{4\pi r}\,sin\,\theta}_{\textit{Element factor}} \underbrace{\int\limits_{-l/2}^{+l/2} I_e(z')\,e^{jkz'cos\theta}\,dz'}_{\textit{Space factor}} \qquad (3.1.1)$$

It can be obtained by an integration of the "elementary source pattern" given in (2.12). It can be noted that the θ component of the far zone field is equal to the product of two patterns, the first is called "Element factor" and the second "Space factor". This characteristic is due to "elementary sources" with the same element factor, independent of the position. Also the far field radiated by a "planar aperture" has the same structure of the "pattern multiplication".

In the case of an Array antenna a mathematical model adopted for the far field is a summation of all the fields of each element:

$$\mathbf{E}(x,y,z) \;=\; \frac{e^{-jkr}}{r}\sum_n \mathbf{G}_n I_n\,e^{j\,\mathbf{k}\cdot\mathbf{r}_n} \qquad (3.1.2)$$

where $\mathbf{G}_n$ characterises the individual element in terms of the polarization and the orientation of the electric field, I_n takes into account the current excitation (in amplitude and phase) of the nth element and $\mathbf{k} = k_x\hat{\mathbf{x}} + k_y\hat{\mathbf{y}} + k_z\hat{\mathbf{z}}$. Only in the case of Arrays with the same type of elements the "element pattern" $G_n = G$ can be factored out and the expression become similar to the continuous case:

$$E(x,y,z) \;=\; \underbrace{\frac{e^{-jkr}}{r}\,G}_{Element\ factor} \; \underbrace{\sum I_n\, e^{j\,\mathbf{k}\cdot\mathbf{r}_n}}_{Array\ factor} \qquad (3.1.3)$$

To the second term, that comes from the spatial distribution of the elements, is attributed the name of "Array factor".

Then, in both cases, continuous or discretized, when the elements are electrically identical and oriented in the same spatial direction, the field can be written as:

Total Field = (Element Factor) x (Space or Array Factor) (3.1.4)

Based on the previous mathematical models, most traditional synthesis methods consists in the *determination of the current distribution*, i.e. $I_e(z')$ or I_n, *so that they radiate the desired Space or Array Factor.* The currents or excitations can be continuous, for aperture antennas, or discretized for array antennas and refer to linear or planar uniform spaced distribution. The fundamental synthesis algorithms are based on the following statements: for *continuous distribution*, when an antenna is modelled by an "Aperture function", the Space Factor is given by an inverse *Fourier Transform* of the antenna current illumination; for *discrete distribution*, the Array Factor of a linear or planar Array Antenna with uniform spaced elements can be seen as an inverse *Discrete Fourier Transform* of the elements excitations or can be written as a *complex polynomial* in terms of its roots.

A possible classification of classical synthesis methods can be derived by the characteristics that the antenna pattern must satisfy. The first category requires that the antenna patterns possess nulls in desired directions. To this end a basic method introduced by

Schelkunoff [22] can be used. The second category, known as "beam shaping", requires that the pattern exhibit a desired distribution. To this class belongs also the "contoured beam" pattern. The Fourier transform and Woodward methods are the principal algorithms based on the "aperture " concept. To the "beam shaping" group belongs also classical methods for the reflector antenna synthesis. These techniques are based on the geometrical optics and the conservation of energy. A large class of antennas requires patterns with narrow beams and low side lobes. These characteristics determine another category of methods. These are the Binomial, the Dolph-Chebycheff and the most used Taylor methods.

A concise description of the basic algorithms of some of previous mentioned methods follows.

Schelkunoff polynomial method: nulls in prefixed directions.

This technique is suitable for the synthesis of an array antenna that requires nulls in prefixed directions. We postulate a linear array of N equally spaced elements located along the **z** axis. The array factor will have no φ variation, but it is a function of the only variable θ:

$$F(\theta) \;=\; \sum_{n=1}^{N} I_n \, e^{j\,(n-1)\,(k\,d\cos\theta - \alpha)} \tag{3.1.5}$$

where I_n are the complex excitation coefficients, d is the inter element spacing and $k = 2\pi / \lambda$. The parameter α can account for a possible uniform progressive phase distribution in the currents. The substitutions $\psi = k\,d\cos\theta - \alpha$ and $w = e^{j\psi}$ convert $F(\theta)$ in:

$$F(w) \;=\; \sum I_n \, e^{j\,(n-1)\,\psi} \;=\; \sum_{n=1}^{N} I_n \, w^{n-1} \tag{3.1.6}$$

The previous polynomial of N-1 degree can be written in terms of its roots

$$F(w) = I_N (w - w_1)(w - w_2)\ldots\ldots(w - w_{N-1}) \tag{3.1.7}$$

The technique consists in choosing suitable roots to determine and control the nulls in the desired directions of the array factor.

Fourier transform method: beam shaping.

The Fourier transform is probably the most used mathematical tool in antenna synthesis.   In this contest it is used to obtain space factor with a prefixed shape.   In a simplest case of a continuous line source of a finite length l the space factor is:

$$SF(\theta) = \int_{-l/2}^{+l/2} I(z) \, e^{\,j\,k\,z\,\cos\theta} \, dz \tag{3.1.8}$$

With the substitution $u = k\,cos\,(\theta)$ and taking into account that the current I(z) is 0 for $z \leq -l/2$ and $z \geq l/2$, we can see that between I(z) and SF there is a relation of Inverse Fourier Transform:

$$SF(u) = \int_{-\infty}^{+\infty} I(z) \, e^{\,j\,u\,z} \, dz \tag{3.1.9}$$

The current will be: $I(z) = \dfrac{1}{2\pi} \displaystyle\int_{-\infty}^{+\infty} SF(u) \, e^{\,-j\,u\,z} \, du.$

Since only source of finite dimension are realizable, the excitation distribution is truncated at $z = \pm l/2$ and the Space Factor will be approximate.

The Woodward method: beam shaping.

The practical application of the properties of the Fourier transform comes to a method due to Woodward.   The analytical formulation of this method is based on the Shannon Sampling Theorem used in communications. The current distribution is decomposed into a sum of uniform amplitude, linear phase sources:

$$I(z) \;=\; \sum_{m=-M}^{M} i_m(z) \;=\; \frac{1}{l} \sum_{m=-M}^{M} a_m(z)\, e^{-j\,k\,z\,\cos\theta_m} \qquad (-l/2 \;\le\; z \;\le\; l/2)$$

$$(3.1.10)$$

θ_m are the angles where the desired pattern will be sampled, $m = \pm 1, \pm 2, \cdots$ (2M even samples) or $m = 0, \pm 1, \pm 2, \cdots$ (2M+1 odd samples). The total reconstructed pattern is obtained by:

$$SF(\theta) \;=\; \sum_{m=-M}^{M} a_m \left\{ \sin\!\left(\frac{k\,l}{2}(\cos\theta - \cos\theta_m) \right) \Big/ \left(\frac{k\,l}{2}(\cos\theta - \cos\theta_m) \right) \right\}$$

$$(3.1.11)$$

The pattern will approximate closely the desired pattern. The $a_m = SF(\theta = \theta_m)$ coefficients are the sampled values of the desired pattern in the $\theta_m = \cos^{-1}(m\,\lambda/l)$ sampling angles.

Dolph Chebyshev method: Narrow beams & low side lobe level.

This method, based on the decomposition of the Array Factor in terms of polynomials, uses the features of Chebyshev polynomial. It oscillates between ± 1 for $-1 \le x \le 1$ (inner region) and increases in absolute value outside. The oscillatory part is mapped onto the "side lobes" and the part of the $x > 1$ region is mapped onto the "main beam" of the pattern. The transformation from the polynomial $T_{N-1}(x)$ to $AF(u)$ is given by $x = x_0 \cos(\pi\,u/2)$, where $u = (l/\lambda)\sin(\theta)$ is the normalized variable and θ is from broadside. Then an N elements array factor corresponds to a Chebyshev polynomial of N-1 degree. The unknown coefficients of the array factor can be determined by equating the series representing the cosine terms of the array factor to the appropriate

Chebyshev polynomial. The resulting Dolph-Chebyshev pattern consists of a main pencil beam, plus side lobes of equal desired levels.

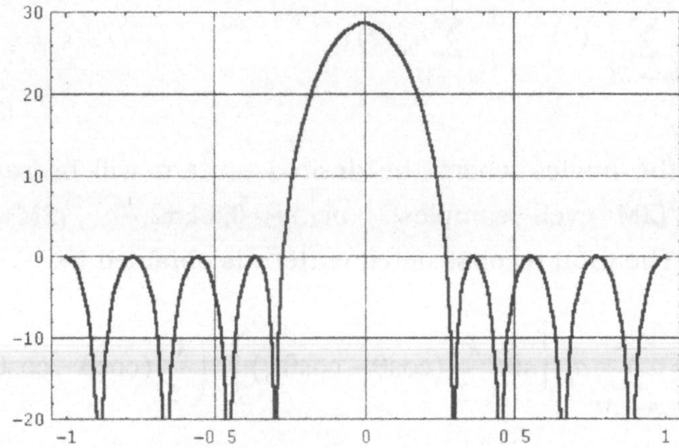

fig. 3.1.1

Dolph-Chebyshev array design yields, for a given side lobe level, the smallest possible first null beamwith.

Taylor distribution method: Narrow beams & low side lobe level.

Taylor design technique [26] yields a pattern that displays an optimum compromise between beamwith and side lobe level. The first few minor-lobes are maintained at an equal and specified level, and remaining lobes decay monotonically. The desired pattern can be written in two forms: a finite product of $\tilde{n}$-1 zeroes or a superimposition of $\tilde{n}$ $sinc(x)$ $sin(x)$ x beams:

$$SF(u) = sinc\,(\pi u)\,\frac{\displaystyle\prod_{n=1}^{\tilde{n}-1}(1 - u^2 / u_n^2)}{\displaystyle\prod_{n=1}^{\tilde{n}-1}(1 - u^2 / n^2)} = \sum_{n=-\tilde{n}+1}^{\tilde{n}-1} SF(n, A, \tilde{n})\, sinc\,\pi(u + n)$$

$$(3.1.12)$$

The zeroes are $z_n = u_n = \pm\sigma\sqrt{A^2 + (n - 1/2)^2}$ for $1 \le n \le \tilde{n}$, $z_n = \pm n$ for $n \ge \tilde{n}$ and the coefficients $SF(n, A, \tilde{n})$ represent samples of the Taylor pattern:

$$SF(n, A, \tilde{n}) = \frac{\left[(\tilde{n} - 1)!\right]^2}{(\tilde{n} - 1 + n)! \; (\tilde{n} - 1 - n)!} \sum_{m=1}^{\tilde{n}-1} (1 - n^2 / u_m^2) \qquad (3.1.13)$$

The scaling factor is $\sigma = \tilde{n} / \sqrt{A^2 + (\tilde{n} - 1/2)^2}$. With the two parameters $\tilde{n}$ and A the designer can control the pattern: $\tilde{n}$ controls the number of sides-lobes at equal level and A their level: $\cosh(\pi A) = desired\_level$.

The linear aperture distribution can be expressed as a finite Fourier series:

$$g(x) = 1 + 2 \sum_{n=1}^{\tilde{n}-1} SF(n, A, \tilde{n}) \cos(n \pi x) \qquad (3.1.14)$$

The following plots show a typical Taylor pattern compared with the pattern from a uniformly excited aperture: $\mathrm{Sin}(\pi u)/(\pi u)$

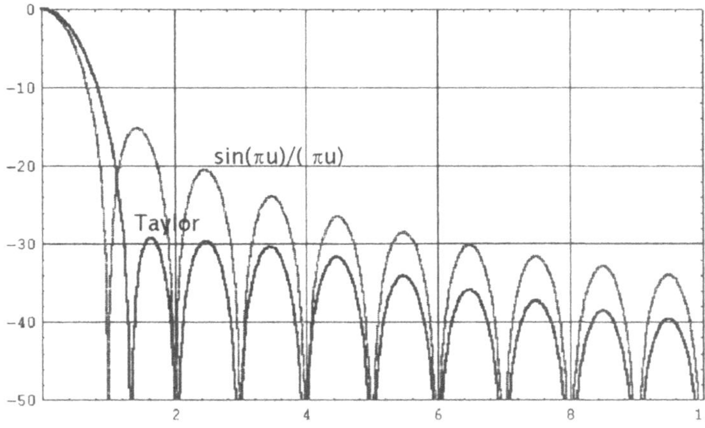

fig. 3.1.2

It's evident the shifting of the first close-in zeroes.

Shaped reflector antennas: beam shaping.

Classical synthesis methods for reflector antennas consist in the design the shape of the reflecting surfaces so that the antenna radiates a far field pattern, called "secondary", with a specified shape. The configuration of the feeds and their overall illumination function, called "primary", is chosen in advance and remain constant during the synthesis procedure.   The shape of secondary pattern can be specified in terms of intensity as the case of antennas for "air traffic control" or in terms of a "contour" that must follow as close as possible, with constant intensity, geographical regions as the case of "satellite applications".

For a single reflector, given a primary illumination $P(\theta)$ and a secondary pattern $S(\theta)$, the determination of the reflector shape $r(\theta)$ is the target for the synthesis.   With reference to fig. 3.1.3 , using Geometrical Optics techniques and the principle of conservation of energy, is possible to determine the function $\varphi = \varphi(\theta)$ that links the incident angle θ to the reflecting φ angle.   Then, with geometrical considerations, the function $\varphi(\theta)$ comes into a differential equation:

$$\frac{dr}{r} = \tan\left[\frac{\theta + \varphi(\theta)}{2}\right] d\theta \qquad (3.1.15)$$

By integration is possible to get the reflector curve $r = r(\theta)$.

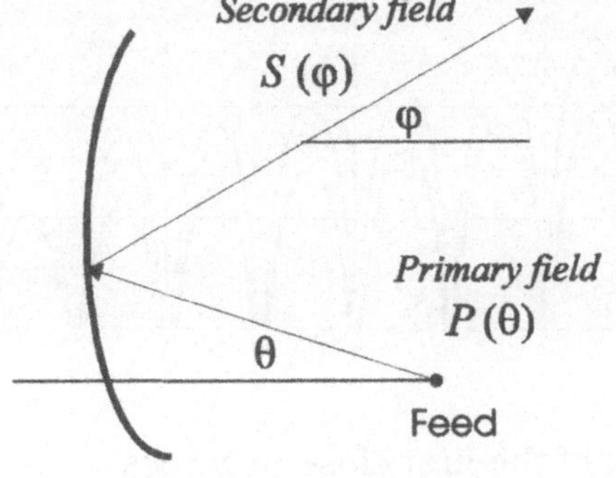

fig. 3.1.3

For the complete tridimensional case, more sophisticated methods formulate the problem as a second order Monge-Amper differential equation:

$$A\, r_{00} + B\, r_{\theta\varphi} + C\, r_{\varphi\varphi} + D\left(r_{\theta\theta}\, r_{\varphi\varphi} + r_{\theta\varphi}^2 \right) + E = 0 \qquad (3.1.16)$$

where A, B, C, D and E are function of $\theta, \varphi, r, r_\theta, r_\varphi$. These techniques have been developed first by Schruben [24] for a single reflector and by Brickell and Wescott [27] for dual reflector antennas.

3.2 Synthesis: a general approach

The present paragraph illustrates a general and unified scheme for the synthesis of radiating system.   It is based on the results of years of scientific collaboration between Alenia company and the Department of Electrical Engineering of University of Naples.   The aim of the study was to settle a general synthesis approach with the following features:

The scheme of the method must be independent of the type of radiating system and must also deal with complex configurations like in the case of "scanning or reconfigurable antennas" and in "conformal array".   The synthesis method must satisfy the complete problem: the determination of both the excitation of sources and their geometrical distribution, i.e. the reflector shape and the primary excitation in a reflector antenna.   Different type of constraints and requirements concerning the far field pattern, the ranges of excitation values and the variability of the antenna geometry, must be simply linked to the mathematical formulation of the problem.   The possibility of using available and consolidate algorithms is also an appreciate characteristic of the approach.

The process leading from the sources to the far field radiated by a generic antenna is described by a scattering operator S: $y = S\ (x, s)$.

It links the primary excitation x to the corresponding radiated far-field y. The operator S depends on the structure of the radiating system characterized by the parameter s.

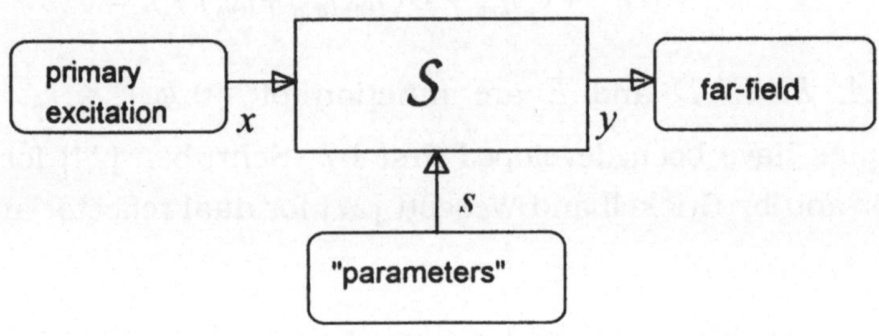

The entities x, y and s belong to appropriate spaces: $x \in X$, X is the space of all excitations; $y \in Y$, Y is the space of all far fields; $s \in S$, S is space of the parameters specifying the geometrical configuration. S is the continuous scattering operator, linear with respect to x, but usually non-linear respect to the parameters s.

The mathematical formulation of the synthesis problem can be the following:

In X, Y and S, assumed to be Hilbert space equipped with a proper mean square norm, the design requirements determine the following subsets:

Y_c: set of far fields complying with the specifications

X_c: set of the excitations satisfying the constraints on the primary source

S_c: set of parameters corresponding to allowable geometry of the radiating system

Given X_c, S_c and Y_c, a solution of the synthesis problem is any couple:

55

$$(x,s) \ \in \ \mathbf{X}_c \times \mathbf{S}_c \ = \ \mathbf{G} \tag{3.2.1}$$

whose image $S(x,s)$ under S belongs to $\mathbf{Y}_c$, i.e. an element of the set:

$$\mathbf{I} \ = \ \mathbf{G} \ \cap \ S^{-1} (\mathbf{Y}_c) \tag{3.2.2}$$

The simple case of a non uniform linear array can clarify the above formulation and focus the difference with the classical method. From eq. (3.1.2) the pattern of a linear array of N elements can be written as

$$\mathbf{E}(u,v) = \sum_n \mathbf{G}_n(u,v) \ I_n \ e^{j\,\mathbf{k}\cdot\mathbf{r}_n} \tag{3.2.3}$$

Where $u = sin\theta \, cos\varphi$ and $v = sin\theta \, sin\varphi$. In this application $\mathbf{X}$ is the space of all N-tuple of complex number C^N, $\mathbf{S}$ is described by all the N-tuple of positions $\mathbf{r}_n$, $\mathbf{Y}$ is the set of all the square-integrable functions and S is the operator definite by (3.2.3).

In order to express the problem in a more operative way and to find an algorithmic scheme for the solution, the concepts of *distance* and *projector* will be introduced.

Given a point z and a subset $\mathbf{W}_c$ of a metric space $\mathbf{W}$: $z \in \mathbf{W}$ and $\mathbf{W}_c \subset \mathbf{W}$, the following operator define the *distance* between z and $\mathbf{W}_c$:

$$d(z,\mathbf{W}_c) = \min_{w \in \mathbf{W}_c} \ \|z - w\| \tag{3.2.4}$$

and the *projector* $\mathcal{P}_{\mathbf{W}_c}$ onto $\mathbf{W}_c$ is the operator which associates the point z to the point nearest to it:

$$\mathcal{P}_{\mathbf{W}_c}: \ z \to \mathcal{P}_{\mathbf{W}_c}(z) \quad \text{such that} \quad \|z - \mathcal{P}_{\mathbf{W}_c}(z)\|^2 \ = \ d^2(z,\mathbf{W}_c) \tag{3.2.5}$$

The necessary and sufficient conditions for (x, s) to belong to $\mathbb{I}$ are:

$$d(x,\mathbf{X}_c) = 0 \qquad d(s,\mathbf{S}_c) = 0 \qquad d(\mathcal{S}(x,s),\mathbf{Y}_c) = 0 \qquad (3.2.6)$$

The sets $\mathbf{X}_c$, $\mathbf{S}_c$ and $\mathbf{Y}_c$ must be closed. In our contest we can always get this condition. When $\mathbb{I}$ is empty, i.e. when the synthesis problem does not have solutions, the relations (3.2.6) can be used to identify the pattern y closest to the set $\mathbf{Y}c$. In other words, a couple (x,s) which minimize $d(\mathcal{S}(x,s),\mathbf{Y}_c)$ under the conditions $d(x,\mathbf{X}_c) = 0$ and $d(s,\mathbf{S}_c) = 0$, is the best choice for the approximate solution.

The synthesis problem is therefore equivalent to the variational problem of minimizing the functional

$$\Phi(x,s) = d^2(\mathcal{S}(x,s),\mathbf{Y}_c) \qquad \text{with} \quad (x,s) \in \mathbf{X}_c \times \mathbf{S}_c \qquad (3.2.7)$$

Using the projection operators, conditions $d(x,\mathbf{X}_c) = 0$ and $d(s,\mathbf{S}_c) = 0$ can be written as $x = \mathcal{P}_{\mathbf{X}_c}(x)$ and $s = \mathcal{P}_{\mathbf{S}_c}(s)$. Our problem becomes the minimization of the functional

$$\Phi(x,s) = \left\| \mathcal{S}\big(\mathcal{P}_{\mathbf{X}_c}(x), \mathcal{P}_{\mathbf{S}_c}(s)\big) - \mathcal{P}_{\mathbf{Y}_c}\big(\mathcal{S}\big(\mathcal{P}_{\mathbf{X}_c}(x), \mathcal{P}_{\mathbf{S}_c}(s)\big)\big) \right\|^2 \qquad (3.2.8)$$

At this point we only need a minimization algorithm providing a sequence $\{x_n, s_n\} = \{g_n\}$ of points of $\mathbf{G} = \mathbf{X}_c \times \mathbf{S}_c$ such that the corresponding sequence $\{\Phi(x_n, s_n)\}$ is not increasing and converging to the minimum of the functional Φ.

When any minimizing sequence is bounded the problem is "weakly well posed". The problem becomes "well posed" if the set $\mathbf{G}$ of the allowable solution is also "finite dimensional". If the number of dimensions or the diameter of $\mathbf{G}$ is too large, the problem become "ill conditioned", i.e. practically "ill posed". When this situation occurs, and this is the case of complex antennas, some kind of *"regularization"* is required. This can be achieved introducing *"a quality criterion"* with a non negative functional $Q(x,s)$ to measure the goodness of the solution: for any $q \geq 0$ the set $\{(x,s): Q(x,s) \leq q\}$ must be bounded. Then, the general formulation of the synthesis problem can be settled as follows.

Given the scattering operator S describing the antenna system; the sets Y_c, X_c and S_c of acceptable far fields, allowable primary excitation and structures; a quality operator $Q(x,s)$, if necessary; find a minimizing sequence for the functional

$$\Psi(x,s) = \Phi(x,s) + \lambda Q(x,s) =$$
$$= \left\| S\left(\mathcal{P}_{X_c}(x), \mathcal{P}_{S_c}(s)\right) - \mathcal{P}_{Y_c}\left(S\left(\mathcal{P}_{X_c}(x), \mathcal{P}_{S_c}(s)\right)\right) \right\|^2 + \lambda Q(x,s)$$

$$(3.2.9)$$

Introducing the expression

$$\mathcal{M}(x, s, y) = \left\| S(x,s) - y \right\|^2 + \lambda Q(x,s)$$

which reduce to Ψ when $y = \mathcal{P}_{Y_c}\left(S\left(\mathcal{P}_{X_c}(x), \mathcal{P}_{S_c}(s)\right)\right)$ and $(x,s) = \left(\mathcal{P}_{X_c}(x), \mathcal{P}_{S_c}(s)\right)$, we can identify the following **iterative scheme**:

$$y^{(n)} = \mathcal{P}_{Y_c}\left(S\left(\mathcal{P}_{X_c}(x), \mathcal{P}_{S_c}(s)\right)\right) \qquad (3.2.10a)$$

$$x^{(n+1)}: \quad \mathcal{M}\left(x^{(n+1)}, s^{(n)}, y^{(n)}\right) = \min_{x \in X_c} \mathcal{M}\left(x, s^{(n)}, y^{(n)}\right) \qquad (3.2.10b)$$

$$s^{(n+1)}: \quad \mathcal{M}\left(x^{(n+1)}, s^{(n+1)}, y^{(n)}\right) = \min_{s \in S_c} \mathcal{M}\left(x^{(n+1)}, s, y^{(n)}\right) \qquad (3.2.10c)$$

It must be noted that the iterative procedure is based on two partial minimizations.

A *fixed geometry synthesis* step: $x^{(n)} \rightarrow x^{(n+1)}$

A *fixed excitation synthesis* step: $s^{(n)} \rightarrow s^{(n+1)}$

It can be shown that the iterative scheme provides a minimizing sequence for Ψ, i.e. $\Psi(x^{(n+1)}, s^{(n+1)}) \leq \Psi(x^{(n)}, s^{(n)})$.

From a computational point of view, some aspects must be underlined. In most cases of relevant interest, simple projectors can be chosen. They can be explicitly evaluated bringing back the functions into the prefixed masks that define the sets. For the minimization steps, the generality of the method allows the use of efficient algorithms or sophisticated minimization techniques. As an example, the complete shaped reflector synthesis can be performed by iterating an aperture field synthesis with a Physical Optics approach and the determination of the reflectors shape by a Geometrical Optics method. The last crucial topic is the choice of the starting point for the iterative procedure. Some suggestions can help the designer: to reduce as much as possible the degree of freedom of the problem; classical and tested pre synthesis can be performed to achieve a starting point sufficiently near to the solution. However, a good experience and sensitivity to the problem is fundamental for an optimal result.

4. Antenna Measurements

The last phase of an antenna project consists in measurements and tests of performances. Today, besides traditional outdoor measurement systems called "Far-Field test range", automatic "Near Field" systems, working in an indoor controlled environment, are used. They allow, in any atmospheric condition, to get much more information in comparison with traditional test-range.

The procedure of the "Near-Field test-range" consists in a **sampling** of the Electric Field "**near**" the antenna, on a prescribed surface, and, with "**Inverse**" algorithms, is possible, numerically, to reconstruct the "**Far Field**" pattern together with the "**Aperture distribution**" that can verify the imposed excitation and eventually defective elements in the antenna. The latter can be considered an "**Identification Problem**", in which the unknown aperture distribution is to be identified from the measured radiated field. Typical near field equipment have "Planar", "Cylindrical" or "Spherical" scanning surface. In this description we concern with the planar scanning because most directive antennas have in the near zone an Electromagnetic Field distribution of finite extent slightly larger than the physical radiating area.

During the last quarter century, the advent of the Fast Fourier Transform (FFT) technique, as well as the rapid progress in electronics, signal processing, and computers, made the "Planar-scanning Near Field Measurement" increasingly fast, accurate, and economical to be implemented. Next paragraph proceeds with the mathematical background of such planar scanning systems, based on the "Plane Wave Spectrum" (PWS) representation of fields.

4.1 Formulation of Antenna Near Fields

In a source-free space region in which near field are measured, the time-harmonic Maxwell equations can be transformed in the following vector wave equations:

$$\nabla^2 \mathbf{E} + k^2 \mathbf{E} = 0 \ ; \qquad \nabla \cdot \mathbf{E} = 0 \qquad\qquad (4.1.1)$$

It can be shown that the following expressions constitute a solution to the above equations for $z \geq 0$ and satisfy the prescribed boundary conditions on the plane $z = 0$.

$$E(x,y,z) = \frac{1}{2\pi} \int\limits_{-\infty}^{+\infty} \int\limits_{-\infty}^{+\infty} A(k_x,k_y)\, e^{\,j\,\mathbf{k}\cdot\mathbf{r}}\, dk_x dk_y \qquad (4.1.2)$$

$$k_x A_x(k_x,k_y) + k_y A_y(k_x,k_y) + k_z A_z(k_x,k_y) = 0 \qquad (4.1.3)$$

where k_x and k_y are real variables, $\mathbf{k} = k_x\hat{\mathbf{x}} + k_y\hat{\mathbf{y}} + k_z\hat{\mathbf{z}}$, $k^2 = \mathbf{k}\cdot\mathbf{k}$, $\mathbf{r} = x\hat{\mathbf{x}} + y\hat{\mathbf{y}} + z\hat{\mathbf{z}}$,

$$\mathbf{A}(k_x,k_y) = A_x(k_x,k_y)\hat{\mathbf{x}} + A_y(k_x,k_y)\hat{\mathbf{y}} + A_z(k_x,k_y)\hat{\mathbf{z}} \qquad (4.1.4)$$

Since the integrand $\mathbf{A}(k_x,k_y)\,e^{\,j\,\mathbf{k}\cdot\mathbf{r}}$ represents a uniform Plane Wave propagating in the $\mathbf{k}$ direction, $\mathbf{A}$ is called the "Plane Wave Spectrum".

In the planar near field measurements, the antenna is placed in the $z \leq 0$ region, as shown in fig. 4.1.1. The scanning plane, specified by $z = z_t$, is conducted on a rectangular grid near the antenna.

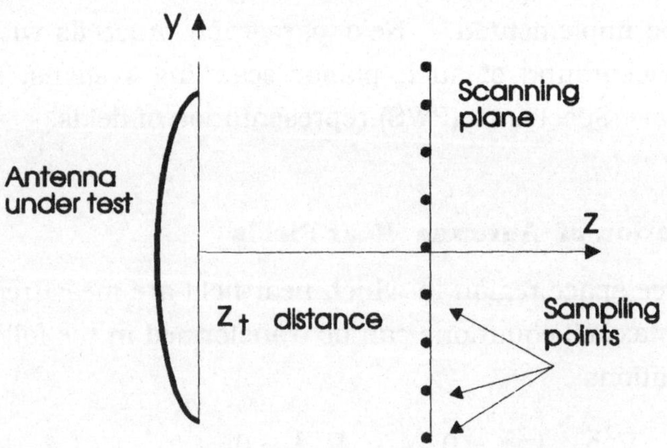

fig. 4.1.1

The radiation condition requires that for $z \geq 0$

$$k_z = \begin{cases} (k^2 - k_x^2 - k_y^2)^{1/2} & \text{if } k_x^2 + k_y^2 \leq k^2 \\ -j(k^2 - k_x^2 - k_y^2)^{1/2} & \text{otherwise} \end{cases}$$

An imaginary k_z corresponds to an evanescent PWS which is rapidly attenuated away from the $z = 0$ plane. Each component of the electric field sampled on the scanning plane is described by

$$E_c(x,y,z_t) = \frac{1}{2\pi} \int_{-\infty}^{+\infty} \int_{-\infty}^{+\infty} A_c(k_x,k_y)\, e^{j\,k_z z_t}\, e^{j\,(k_x x + k_y y)}\, dk_x dk_y \qquad (4.1.5)$$

where the subscript c indicates the x, y or z component. It is evident that the previous equation is an inverse **Fourier Transform**, then the plane wave spectrum $A_c(k_x,k_y)$ can be extracted by means of a two dimensional transform and taking into account the phase $e^{j\,k_z z_t}$ due to the distance z_t.

The PWS is directly related to the desired Far Field pattern. In fact a simple and useful relationship between the field pattern and the PWS in the far zone of the antenna is

$$\mathbf{E}(x,y,z) = \frac{j e^{j\,kr}}{r}\, k_z\, \mathbf{A}(k_x,k_y) \qquad (4.1.6)$$

where: r is the distance in a spherical coordinate system centred in $x = y = z = 0$, k_z is always real because an imaginary value corresponds to an evanescent PWS that does not propagate to the far zone, and the expressions for k_x, k_y, k_z are

$$k_y = k\,\sin\theta\,\sin\phi\,, \quad k_x = k\,\sin\theta\,\cos\phi\,, \quad k_z = k\,\cos\theta$$

To obtain the "Field Aperture" distribution, that is one of the most important designing function for an antenna, it is necessary a direct Fourier transform as indicated in the following expression

$$E_c(x,y,0) = \frac{1}{2\pi} \int_{-\infty}^{+\infty} \int_{-\infty}^{+\infty} A_c(k_x,k_y)\, e^{j\,(k_x x + k_y y)}\, dk_x dk_y \qquad (4.1.7)$$

In this paragraphs basic mathematical concepts have been outlined, in the following, a more realistic application will be described.

4.2 Near-Field test range as a Diagnostic Tool in an Array Antenna

In actual Planar Near-Field Measurement (PNFM) some other entities come into a more complex formulation. The most important function that must be taken into account is the PWS of the electromagnetic probe, $\mathbf{S}(k_x,k_y)$, used to sample the field on the planar surface. When the antenna under test is a planar array of radiating elements of the same type, the element pattern, $\mathbf{g}(k_x,k_y)$, must be also considered.

This paragraph concern the "inverse" procedure used to obtain the array excitation $F(x_m,y_n)$ from which is possible to get information about the quality of the feeding structure.

Before starting with the measure some preliminary steps must be done. The *choices* of: the *distance z_l* for the scanning plane, often it is few wavelength from the antenna aperture; the *extension of the scanning area* and the *sampling intervals Δx* and Δy of the rectangular grid, these decisions are made with the help of the sampling theory; the *type of the probe*, that is itself a small antenna, and his PWS $\mathbf{S}(k_x,k_y)$ must be completely known. The probe output signal $B_0(k_x,k_y)$ at a given point is a complex function with an amplitude and phase component, and can be viewed as the contribution of all the plane wave components received:

$$B_0(x,y,z_l) = \frac{1}{2\pi} \int_{-\infty}^{+\infty} \int_{-\infty}^{+\infty} \mathbf{T}(k_x,k_y) \cdot \mathbf{S}(k_x,k_y)\, e^{j\,k_z z_l}\, e^{j\,(k_x x + k_y y)}\, dk_x dk_y$$

$$(4.2.1)$$

where $\mathbf{T}(k_x,k_y)$ is the PWS of the antenna under test.

A 2D Fourier transform of $B_0(x,y,z_t)$ gives

$$D(k_x,k_y) = \mathbf{T}(k_x,k_y)\cdot\mathbf{S}(k_x,k_y)\,e^{j\,k_z z_t} =$$
$$= \frac{1}{2\pi}\int_{-\infty}^{+\infty}\int_{-\infty}^{+\infty} B_0(x,y,z_t)\,e^{-j\,(k_x x + k_y y)}\,dk_x dk_y \qquad (4.2.2)$$

From this expression it evident that the far field spectrum $\mathbf{T}(k_x,k_y)$ can be extracted from $D(k_x,k_y)$ by removing the probe response $\mathbf{S}(k_x,k_y)$. This step of the procedure is usually referred as the "**Probe Correction**". With the far field spectrum $\mathbf{T}(k_x,k_y)$ determined, one can proceed to recover the discrete aperture distribution. The array excitation $F(x_m,y_n)$ is related to the far field spectrum by

$$\mathbf{T}(k_x,k_y) = \mathbf{g}(k_x,k_y)\sum_{m=1}^{M}\sum_{n=1}^{N} F(x_m,y_n)\,e^{-j\,(k_x x_m + k_y y_n)}\,\Delta x \Delta y \qquad (4.2.3)$$

At this point, after eliminating the element pattern $\mathbf{g}(k_x,k_y)$ from (4.2.3), with the step referred to as "**Element Correction**", one can take a discrete inverse transform to obtain $F(x_m,y_n)$.

5 References

[1] C. A. Balanis: *Antenna Theory: Analysis and Design.* New York: Wiley, 1982

[2] C. A. Balanis: *Advanced Engineering Electromagnetics.* New York: Wiley, 1989

[3] O. M. Bucci, G. Franceschetti, G. Mazzarella, and G. Panariello: *A general projection approach to array synthesis.* 1989 Antennas Propag. Symp., San José 1989

[4]   O. M. Bucci, and G. D'Elia: *A general and effective approach to the synthesis of shaped reflector antennas.*   1989 Antennas Propag. Symp., San José 1989

[5]   O. M. Bucci, G. D'Elia, and G. Leone: *Reflector antenna power synthesis: a general and efficient approach.*   IEEE Trans. Antennas and Propagat., AP-37, pp. 875-883, 1989

[6]   O. M. Bucci, G. Franceschetti, G. Mazzarella, and G. Panariello: *The intersection approach to array synthesis.*   Proc. IEE pt. H, 137, pp. 349-357, 1990

[7]   O. M. Bucci, and G. D'Elia: *A new approach to the synthesis of scanning and reconfigurable beam reflector antennas.*   XXIII General Assembly of the URSI Symposium, Prague, Czechoslovakia,1990

[8]   O. M. Bucci: *Reflector antenna synthesis: a general framework and solution algorithm.*   2nd Int. Conf. on Electromagnetics in Aereospace Appl., Torino 1991

[9]   C. L. Dolph: *A current distribution for broadside arrays which optimizes the relationship between beamwith and side lobe level.* Proc. IRE, vol. 34, pp.335-348, 1946

[10] R. S. Elliott: *Antenna Theory and Design.*   New York: Prentice Hall, 1981

[11] R. S. Elliot: *Design of line-souce antennas for sum patterns with sidelobes of individually arbitrary heights.*   IEEE Trans. Antennas Propagat., vol. AP-24, pp. 76-83, Jan 1976

[12] J. D. Hanfling, G. U. Borgiotti, and L. Kaplan: *The backward transform of the near field for reconstruction of aperture fields.* Proc. IEEE Antennas and Propagat. Symp., vol. 2, pp. 764-767,1979

[13] R. F. Harrington: *Field Computation by Moment Method.* New York Macmillan, 1968

[14] J. B. Keller: *Diffraction by an Aperture.*   J. Appl. Phys., vol. 28, pp. 426-444, Apr. 1957

[15] J. B. Keller: *Geometrical Theory of Diffraction.*   J. Opt. Soc. Amer., vol. 52, pp. 116-130, Feb. 1962

[16] R. G. Kouyoumijan and P. H. Pathak: A Uniform *Geometrical Theory of Diffraction by an edge in a perfectly conducting surface.* Proc. IEEE., vol. 62, pp. 1448-1461, Nov. 1974

[17] J. D. Kraus: *Antennas.*   New York: McGraw Hill, 1952

[18] A. Levi and H. Stark: *Image restoration by the method of generalized projections with applicatioon to restoration from magnitude.*   J. Opt. Soc. Am pt. A, A1, pp. 932-943, 1984

[19] A. C. Newell: *Planar Near-Field Measurements.*   Nat. Bur. of Stand., June 1985

[20] G. T. Poulton: *Antenna power synthesis using method of successive projections.*   Electron. Lett., 22, pp. 1042-1043, 1986

[21] S. A. Schelkunoff: *Some Equivalence Theorems of Electromagnetis and their Application to Radiation Problems.*   Bell Syst. Tech. J., vol. 15, pp. 92-112, 1936

[22] S. A. Schelkunoff: *A mathematical theory of linear arrays.*   Bell Syst. Tech. J., vol. 22, pp. 80-107, Jan 1943

[23] S. A. Schelkunoff: *Advanced Antenna Theory.*   New York: Wiley, 1950

[24] J. S. Schruben: *Formulation of a reflector-design problem for a lighting fixture.*   J. Opt. Soc., vol. 62, No. 12, Dec 1972

[25] W. L. Stutzman and G. A. Thiele: *Antenna Theory and Design.* New York: Wiley, 1981

[26] T. T. Taylor: *Design of line-souce antennas for narrow beamwith and low sidelobes.*   IRE Trans. Antennas Propagat., vol. AP-7, pp.16-28, Jan 1955

[27] B. S. Wescott, F. A. Stevens, and F. Brickell: *GO synthesis of offset dual reflectors.*   Proc. Inst. Elec. Eng., vol. 128, pt. H. no. 1, Feb. 1981

[28] P. M. Woodward: *A method of calculating the field over a plane aperture required to produce a given polar diagram.*   Proc. Inst. Elect. Eng., pp. 1554-1558, Jan 1947

Ugo F. D'Elia

Alenia S.p.A.

via Tiburtina km. 12.400

00131 Rome (Italy)

Parameter Identification Problems in Photographic Science

Kam-Chuen Ng   David S. Ross
Research Laboratories, Eastman Kodak Company

Abstract

In this paper, we present a reaction-diffusion model of the development of photographic film. We discuss some of the chemistry and physics of the photographic process, and we discuss how these are captured in the model. For some simplifications of the model we formulate and discuss the inverse problem of determining parameters from experimental data.

INTRODUCTION

Color photographic film consists of several thin chemically active layers of photographic emulsion coated on a base (Fig.1.) The layers are composed essentially of silver halide crystals and oil droplets suspended in gelatin. The different layers capture light of different colors, and at different 'speeds'.

The silver halide crystals – most commonly silver bromide or silver bromo-iodide – have dimensions on the order of 1 micron. They come in a variety of shapes – e.g. cubic, octahedral and tabular, that are used for different purposes. Their function is to capture light when the film is exposed. Grains are coated with sensitizing dyes to allow them to capture light of the desired color. The process of light capture is complex and incompletely understood, and its details are beyond the scope of this paper. The interested reader might consult [1], chapter 14 of [2] or the pertinent chapters in [3]. (The book by James [4] is the definitive volume on photographic science. The book by Carroll, Higgins and James [3] is somewhat less detailed and more elementary. The survey by Thirtle and Zwick [5] is even more accessible). For our purposes, this summary should suffice; grains that are struck by enough photons on exposure form *latent image sites*. The fraction of grains, in any region of the film, that have formed such sites is a function of the intensity of the incident light of the appropriate color.

The oil droplets are spherical, with diameters on the order of .1 micron. They contain chemicals – *couplers* - that form colored dyes and other chemicals during the development process. These dyes constitute the visible image that is produced during development.

In development, exposed film is immersed in an aqueous solution. The gelatin swells as it incorporates water. *Reduced developer*, a chemical that diffuses into the film and reacts with silver halide grains at latent image sites. The developer transfers electrons to the silver halide grain, i.e. the developer is oxidized. The electrons combine with interstitial silver ions to form pure silver (in black and white film, there are no dyes, the image is composed of the pure silver). The oxidized developer diffuses away from the grain and reacts with the couplers in the oil droplets to form dye and *inhibitor*. The inhibitor diffuses, and can adsorb to the surface of silver halide grains, where it acts to slow down the development. A schematic depiction of this process is shown in Fig. 2. A mathematical discussion of the purpose and mechanisms of inhibitors is presented in chapter 10 of [2].

This is a very abbreviated description of photographic development. We have presented just an outline of the essentials, sufficient for formulating the mathematics problems that are the topic of this paper. In reality, there are chemical species that we have not mentioned (e.g. sulfite) that play important roles in the process. And the processes that we have discussed, particularly the reduction of silver halide to elemental silver, are more complex than our brief outline might suggest.

MODELING

Our model will be expressed as a system of reaction-diffusion equations for the following functions,

$R(x,t)$ Concentration of reduced developer, in moles/cc
$X(x,t)$ Concentration of oxidized developer, in moles/cc
$C_1(x,t)$ Concentration of image coupler, in moles/cc

$C_2(x,t)$ Concentration of DIR (Development Inhibitor Releasing) coupler, in moles/cc

$D(x,t)$ Concentration of dye, in moles/cc

$P(x,t)$ Concentration of inhibitor, in moles/cc

$P^*(x,t)$ Concentration of adsorber inhibitor, in moles/cc

$S(x,t)$ Concentration of free silver halide surface, in cm^2/cc

Here, t is the time, in seconds, that has passed since the film was immersed in the development solution, and x is the depth, in cm, into the film. In many applications, for example the study of edge enhancement and acutance [2, chapter 10], these concentrations would depend on another spatial coordinate, one parallel to the surface of the film, to account for spatial variations in exposure. Here, we consider only uniform exposures, we are taking pictures of uniformly painted and illuminated walls.

In addition, we need the following parameters,

D_R Diffusion coefficient of reduced developer, in cm^2/sec

D_R Diffusion coefficient of oxidized developer, in cm^2/sec

D_R Diffusion coefficient of inhibitor, in cm^2/sec

K_1 Reaction rate of Oxidized Developer and Image Coupler, in cc/(mole-sec)

K_2 Reaction rate of Oxidized Developer and DIR Coupler, in cc/(mole-sec)

K_3 Adsorption rate of inhibitor, in cc/(cm^2-sec)

K_4 Reaction rate of Oxidized Developer and DIR Coupler, in sec$^-1$

λ Surface area occupied by a mole of adsorbed inhibitor particles, in cm^2/mole

μ Molar volume of silver halide, in cc/mole

These parameters are all *effective* parameters; they vary spatially because of the inhomogeneity of the medium. Reduced developer, say, diffuses at one rate in aqueous gelatin, at another rate in oil drops, and at still another in silver halide. The reaction between oxidized developer and the couplers occurs at one rate in the oil drops, and does not occur elsewhere, as there is no coupler elsewhere. Our formulation will treat the medium, the photographic emulsion, as if it were homogeneous. Some discussion of the application of mathematical homogenization theory to this problem is presented in [2, chapter 10].

In problems in which there are two or more layers, the parameters will vary from layer to layer. For example, diffusion will be faster in pure gelatin spacer layers than in emulsion layers [6]. Here, we shall treat only a single layer; the extension to the more general case is straightforward.

Reduced developer diffuses in from solution, and is consumed in the oxidation reaction with exposed silver halide grains. Its equation is

$$\frac{\partial R}{\partial t} = D_R \frac{\partial^2 R}{\partial x^2} - f(R, X, S, P^*)$$

Here, the function $f(R, X, S, P^*)$ describes the development rate. The detailed form of this function is not essential for our purposes, and it is complicated. An elementary discussion is presented in [2, chap.10]. Much more detailed presentations can be found in [7] and the relevant articles in [3] and [4]. This function is dependent on the fraction of developable grains, of course, and it has the property that

$$\frac{\partial f}{\partial P^*} < 0$$

The development rate decreases as the concentration of adsorbed inhibitor increases.

Oxidized developer is produced in the reaction in which reduced developer is consumed. It also diffuses, and it reacts with couplers. It undergoes other reactions (e.g., it is 'scavenged' by sulfite), which we omit;

$$\frac{\partial X}{\partial t} = D_X \frac{\partial^2 X}{\partial x^2} + f(R, X, S, P^*) - K_1 X C_1 - K_2 X C_2$$

The equations for the couplers are simple, they are both immobile (because they are fixed in oil droplets which are fixed in the gelatin matrix) and are consumed in reactions with oxidized developer,

$$\frac{\partial C_1}{\partial t} = -K_1 X C_1$$

$$\frac{\partial C_2}{\partial t} = -K_2 X C_2$$

Dye is produced in the reaction of oxidized developer with image coupler, and it is immobile,

$$\frac{\partial D}{\partial t} = K_1 X C_1$$

Inhibitor is produced in the reaction of oxidized developer with DIR coupler. it diffuses, and is lost through adsorption to silver halide grains. This adsorption is modeled as a second-order reaction between the inhibitor and free silver halide surface. Inhibitor is 'produced' when adsorbed inhibitor desorbs from the silver halide surface. This desorption is modeled as a first order reaction,

$$\frac{\partial P}{\partial t} = D_P \frac{\partial^2 P}{\partial x^2} + K_2 X C_2 - K_3 P S + K_4 P^*$$

Adsorbed inhibitor is immobile, and it is exchanged with inhibitor through the adsorption and desorption processes that we just described,

$$\frac{\partial P}{\partial t} = K_3 P S - K_4 P^*$$

The free silver surface decreases as development proceeds. It also decreases as inhibitor adsorbs to it, and increases when adsorbed inhibitor desorbs. Free silver surface has different units from the other species, so its equation contains factors to account for this. Also, it contains a geometry factor that multiplies the development rate. We have used the geometry factor for cubic crystals,

$$\frac{\partial S}{\partial t} = -\left(\frac{2}{3}6^{\frac{2}{3}}\right) \mu S^{-\frac{1}{2}} f(R, X, S, P^*) - \lambda(K_3 P S - K_4 P^*)$$

The model that we have defined here is a conceptual intermediary between models that are used in practice. It is more complex than is necessary for parameter identification, and it is too simple as a model of the actual development process. It does capture the essence of the process, however, we hope that it provides a sufficient background for the parameter identification problems to which we now turn.

PARAMETER IDENTIFICATION

In order to use models of the type that we discussed in the previous section, we must have values for the diffusion parameters and the reaction constants. To obtain these parameters, we use simplified experimental films and analogously simplified models. For instance, we might place a single emulsion layer containing coupler in contact with a tank containing a reducing agent (oxidized developer) to measure the coupling reaction rate K_1 or K_2 or the diffusion coefficient D_X. Or we might have a similar arrangement with inhibitor in the tank to determine D_P, K_3 and K_4. Some specific cases are reported by Zhao et. al. in [9].

As a general case, we consider the experimental set-up depicted in Fig. 3; a tank containing a chemical A is placed in contact with a three-layer film, the central layer of which contains chemical B. Species A is mobile, it diffuses from the tank into the film. Species B is immobile. The two species react in a second-order reaction to create species C. We shall consider this reaction to be irreversible; the generalization to reversible reactions is straightforward, but it does introduce another parameter to be fitted.

If we denote the concentrations of the species by $A(x,t)$ and $B(x,t)$, we have the equations

$$\frac{\partial A}{\partial t} = D\frac{\partial^2 A}{\partial x^2} - KAB$$
$$\frac{\partial B}{\partial t} = -KAB$$

with the boundary conditions

$$\left.\frac{\partial A}{\partial x}\right|_{x=L} = 0$$

$$\left.\frac{\partial A}{\partial t}\right|_{x=0} = \left.\frac{\alpha D}{V}\frac{\partial A}{\partial x}\right|_{x=0}$$

The first condition is just a no-flux condition, indicating that the backing of the film is impenetrable. In the second condition, α is the cross-sectional area of the tank-film interface, and V is the volume of the tank. The second condition follows from the assumption that the tank is well-stirred, so that the concentration of A throughout the tank is equal to its concentration at the tank-film interface; the condition says that the rate at which A is depleted from the tank is equal to the rate at which it flows into the film. There are no boundary conditions for B, since it is immobile.

We also need initial conditions,

$$A(x,0) = \begin{cases} 0 & 0 < x \le L \\ A_0 & x = 0 \end{cases}$$

$$B(x,0) = \begin{cases} B_0 & L_1 < x \le L_2 \\ 0 & \text{otherwise} \end{cases}$$

These conditions say that B initially has uniform concentration B_0 in the middle layer and is absent from the other layers, and A initially has concentration A_0 in the tank, and has not yet diffused into the film. In practice, there are two types of inverse problems that we need to solve:

Problem 1: Given $A(0,t)$, determine D and K.

This problem corresponds to an experiment in which the concentration of A in the tank is monitored – say, by and optical method, as a function of time.

Problem 2: Given $\int_0^L B(x,t)dx$, determine D and K.

This problem corresponds to an experiment in which a series of identical films is used. Each film is removed from the tank after a certain time, and the total amount of B remaining after that time is measured.

In each of these problems, we have assumed for simplicity that the diffusion coefficient is the same in all layers, which may not be the case [6].

Existence, uniqueness, and regularity of solutions of the forward problem are discussed by Friedman and Reitich in [10]. These authors also prove that the inverse problem is well-posed, that the data given in problem 1 or problem 2 is sufficient to determine D and K.

In practice,these problems are solved numerically by doing a nonlinear least-squares fit of data to solutions generated by finite difference methods. For example, for problem 1, the parameters and are chosen to minimize the objective function

$$\phi(D, K) = \frac{1}{2} \sum_j W(t_j)(A_N(t_j) - A_E(t_j))^2$$

Here, $A_E(t_j)$ is the experimental value of the concentration in the tank at time t_j, $A_N(t_j)$ is the value of the numerical solution at the same time, and $W(t)$ is a weight function. We should like to choose $W(t)$ to make the problem as well-conditioned as possible.

Our purpose here has simply been to present this problem. However, we shall conclude with an interesting observation we have made in our numerical work on the problem, which may serve as a starting point for theoretical investigations.

The condition of the inverse problem is determined by the partial derivatives of the objective function,

$$\frac{\partial \phi}{\partial D} = \sum_j W(t_j)(A_N(t_j) - A_E(t_j))\frac{\partial A_N(t_j)}{\partial D}$$

$$\frac{\partial \phi}{\partial K} = \sum_j W(t_j)(A_N(t_j) - A_E(t_j))\frac{\partial A_N(t_j)}{\partial K}$$

These derivative can be computed numerically by adding to the system of equations (1) the partial differential equations and boundary conditions for $\frac{\partial A}{\partial D}$, $\frac{\partial A}{\partial K}$, $\frac{\partial B}{\partial D}$, and $\frac{\partial B}{\partial K}$, which are obtained by differentiating the equations and initial and boundary conditions in (1) with respect to D and K. For example,

$$\frac{\partial}{\partial t}\left(\frac{\partial A}{\partial D}\right) = \frac{\partial^2 A}{\partial x^2} + D\frac{\partial^2}{\partial x^2}\left(\frac{\partial A}{\partial D}\right) - \left(K\frac{\partial A}{\partial D}B + KA\frac{\partial B}{\partial D}\right)$$

$$\frac{\partial}{\partial x}\left(\frac{\partial A}{\partial D}\right)\bigg|_{x=L} = 0$$

$$\frac{\partial}{\partial t}\left(\frac{\partial A}{\partial D}\right)\bigg|_{x=0} = \frac{\alpha}{V}\frac{\partial A}{\partial x}\bigg|_{x=0} + \frac{\alpha D}{V}\frac{\partial}{\partial x}\left(\frac{\partial A}{\partial D}\right)\bigg|_{x=0}$$

$$\frac{\partial A}{\partial D}\bigg|_{t=0} = 0$$

Graphs of the functions $\frac{A(0,t)}{A_0}$, $\frac{\partial A(0,t)}{\partial D}\frac{D}{A_0}$ and $\frac{\partial A(0,t)}{\partial K}\frac{K}{A_0}$ for a particular choice of parameters are shown in Fig. 4. Note that the derivates have extrema; there is a 'best' time interval in which to take data to determine the parameters. The weighting function should be chosen to give greater weight to points near the extrema.

References

[1] Bayer, B. E.; Hamilton, J. F:, Computer investigation of a latent image model, J. Optical Society, 55, 4, (1965).

[2] Friedman, A., Mathematics in Industrial Problems, Part 2, Springer-Verlag, New York, 1988.

[3] Carroll, B. H.; Higgins, G. C.;James, T. H.: Introduction to Photographic Theory, Wiley-Interscience, New York,1980.

[4] James, T. H.: The Theory of the Photographic Process, 4th. ed., Macmillan, New York, 1977. Article by James, 373-403

[5] Thirtle, J. R.; Zwick, D. M: Color Photography, in Kirk-Othmer Encyclopedia of Chemical Technology, Vol 5., 1964, pp. 812-845.

[6] Ng, K. C.; Ross, D. S.: Diffusion in Swelling Gelatin. J. Imaging. Sci., 35 (1991) 356-361.

[7] Matejec, R.; Meyer, R.: Contribution on the Mechanism of Photographic Development. Photogr. Sci. Eng., 7, 265 (1963).

[8] Xia, P. J.; Zhao, W. F.; Ren, X. M: Kinetics Modeling of Coupling Reactions in Gelatin Layers. J. Imaging. Sci., 37 (1993) 354-358.

[9] Zhao, W. F.; Mo, Y. B.; Xia, P. J.; Ren, X. M: Kinetic of Coupling Reactions of Solvent Dispersions in Gelatin Layers. J. Imaging. Sci., 37 (1993) 354-358.

[10] Friedman, A.; Reitich, F: Parameter identification in reaction-diffusion models. Inverse Problems, 8, 187 (1992).

David S. Ross
Kodak Research Laboratories
Rochester, NY 14650   USA

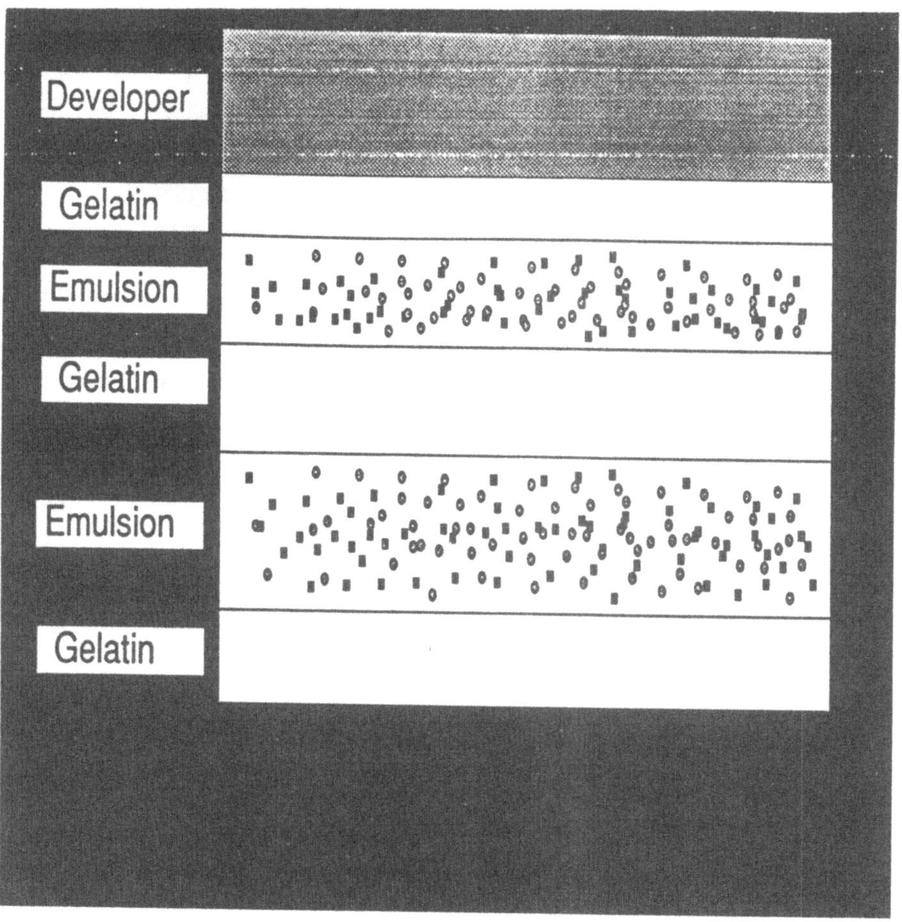

Figure 1: Photographic Emulsion

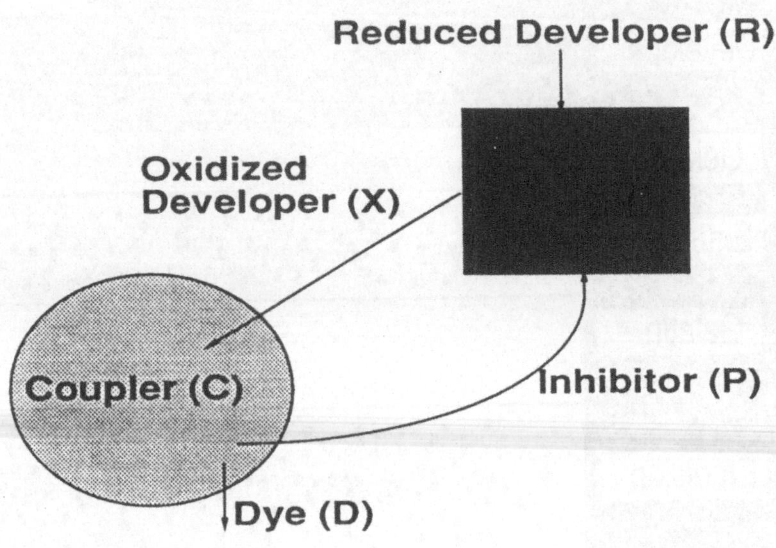

Figure 2: A schematic depiction of developing a film

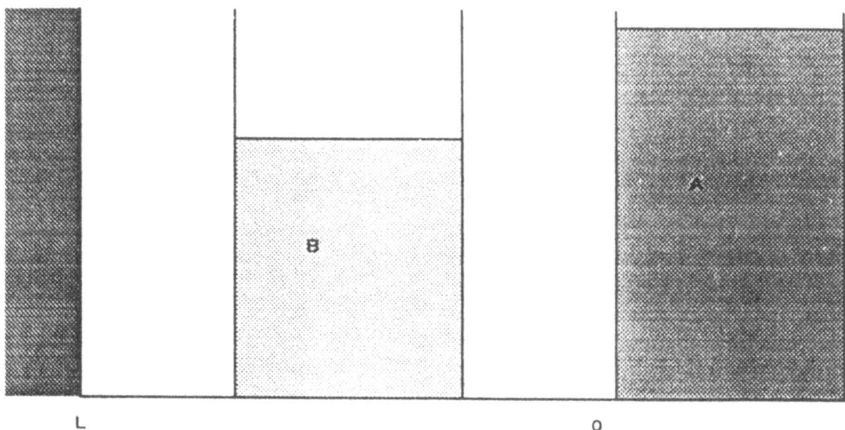

Figure 3: Experimental Set-up for parameter identification

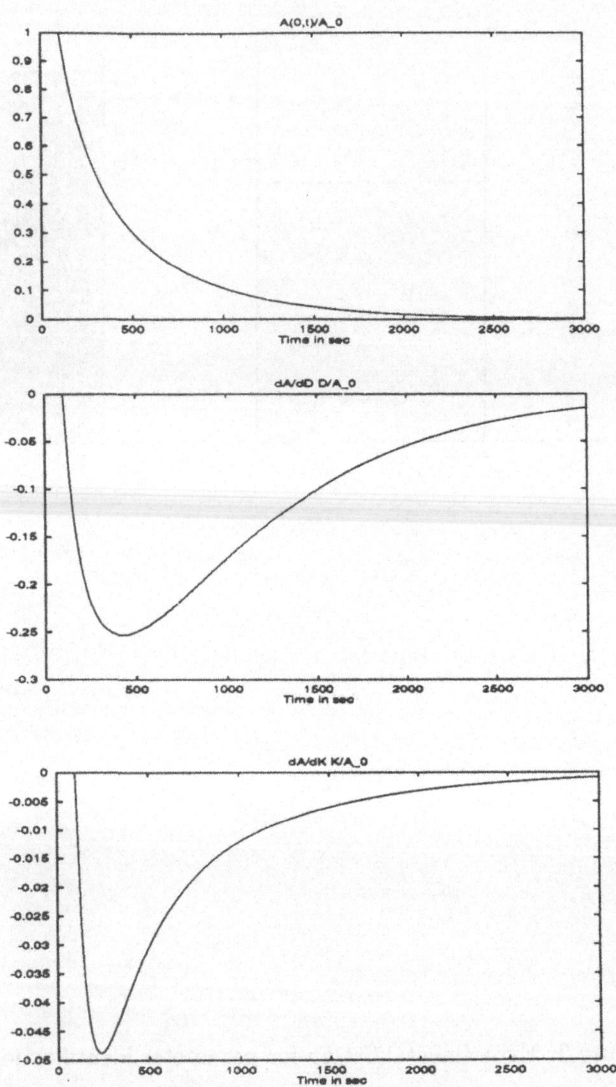

Figure 4: Graphs of $\frac{A(0,t)}{A_0}$, $\frac{\partial A(0,t)}{\partial D}\frac{D}{A_0}$ and $\frac{\partial A(0,t)}{\partial K}\frac{K}{A_0}$ for $D = 2.3 \times 10^6\,\text{cm}^2/\text{sec}$, $K = 2.3 \times 10^4\,\text{litre}/(\text{moles sec})$, $A_0 = 1.5 \times 10^6\,\text{moles/litre}$ $B_0 = 3.7 \times 10^3\,\text{moles/litre}$, $L_1 = 2\mu\text{m}$, $L_2 = 12\mu\text{m}$, $L = 14\mu\text{m}$ and $\frac{\alpha}{V} = 0.746\text{cm}^{-1}$

PARAMETER AND SYSTEM IDENTIFICATION FOR FLUID FLOW IN UNDERGROUND RESERVOIRS

A.T. Watson
Department of Chemical Engineering
Texas A&M University
College Station, TX 77843-3122

J.G. Wade
Department Mathematics and Statistics
Bowling Green State University
Bowling Green, OH 43403

R.E. Ewing
Institute for Scientific Computation
Texas A&M University
College Station, TX 77843-3404

1 Introduction

The mathematical modeling and simulation of the flow of fluids in underground reservoirs is an essential exercise for planning aspects associated with two seemingly disparate applications: production of petroleum and the remediation of water resources. Mathematical simulation is used in order to: (1) evaluate different recovery (or remediation) processes, (2) help make decisions regarding the placement of wells, and (3) specify various operating procedures associated with injection and production wells.

Equations to describe the flow of fluids in porous media are typically based on a continuum (or macroscopic) representation, in which state variables are defined relative to representative volume elements, or local volume averages [1, 37, 42]. Several porous media properties are defined within these equations. These properties are typically functions of location, and may also depend on state variables, such as fluid saturations. While the equations for modeling many different displacement processes are fairly well-established, the specification of the appropriate porous media properties to use in simulating flow behavior for specific situations remains an enormous problem. The difficulty of this problem is in part due to the spatial dependence of the properties, and the inaccessibility of the reservoir to measurements. Another feature is that many of the properties are defined within the constitutive equations for flow, and are not directly measureable. The problem of property estimation (or "reservoir characterization") is one of the key elements for success of the modeling exercise. (Here and throughout, the terms "rock" and "reservoir" are used in a generic sense; "rock" may include soil in the unsaturated zone in hydology applications, and "reservoir" refers to whatever geological region or formation comprises the porous medium of interest in a given application.) See [14, 20, 21, 39] for surveys of parameter estimation techniques for porous media applications.

Seismic data and other sorts of geological studies are typically used in defining reservoir structure and boundaries, but usually provide little information about the various rock properties. Rock samples, or cores, which have been extracted from the reservoirs can provide some localized information. Estimates of rock properties can be obtained from laboratory experiments conducted on plugs removed from the core samples. However, core sample extraction is a costly exercise, so relatively few locations will be sampled, and the spatially variable properties will remain poorly determined. Furthermore, care should be taken in the use of such information since the values obtained represent a different scale than is normally used in representing the macroscopic flow properties for simulating field-scale performance. Once wells have been drilled, information about rock properties can be obtained from well logs, which can provide measures of various properties in the vicinity of wellbores. However, this information is also of limited value. Generally, well logs can provide good information regarding the porosity and fluid saturations in the vicinity of the wells, but typically provide little information about properties associated with fluid transport, such as absolute and relative permeabilities.

Monitoring of flow at wells in the reservoir can provide the best source of information about the properties required for simulation. Such data can include pressure and flow rates of various fluid phases obtained during production, or during field-scale experiments, such as well or tracer tests. The estimation of rock properties on the basis of these data can be posed as an inverse problem. Unlike the other data, suitable solutions of this problem can provide information about properties at locations away from the wells and with a scale that is suitable for the field-scale simulations.

The inverse problem of estimating porous media properties from measured flow behavior is considered here. Much work has been directed to the associated parameter estimation problem–determination of parameters within functional representations of properties from measured data. This problem is generally addressed through the mathematical minimization of a least-squares objective functional. Significant progress has been made with this problem. With the introduction of variational methods [9, 11], efficient quasi-Newton algorithms [57], and the use of inequality constraints [57, 58], as well as the fantastic advances in computer hardware, relatively large-scale minimization problems can now be handled fairly efficiently. Now it is particularly important to address other aspects associated with these inverse problems in order to provide effective solutions for situations of industrial importance.

A key feature of these inverse problems is that, due to the relatively large dimension of the unknown parameter space, and the relatively small number and limited quality of data, they are typically *ill-posed* as discussed below. While one can strive to increase the amount and quality of data, it will always be necessary to suitably reduce the unknown parameter space if meaningful and useful results are to be obtained.

In this article, work addressing the ill-posedness aspects of the inverse problems associated with fluid flow in underground reservoirs is discussed. In §2 a description of the general class of models under consideration is presented, and basic features of the porous media properties and estimation of those properties are discussed. General considerations of computational aspects of associated parameter estimation problems are summarized in §3. In the next section, the general approaches used to address the ill-posedness aspects are discussed. Examples of applications of some of these methods to industrial problems are presented.

2 Model Equations and Inverse Problems

In this section a general model for the immiscible flow of multiple fluid phases is presented. This model and various adaptations have important petroleum and hydrology applications. For example, it is often the case that oil is found in geological formations together with water and/or natural gas, the presence of which must be accounted for in the modeling process. Moreover, improved oil recovery techniques usually involve the injection of fluids such as water into the reservoir with the goals of maintaining reservoir pressure and sweeping the oil to production wells. Also, popular models of groundwater flow (e.g., Richards equations [1]) are based on certain simplifications of the multiphase model. For more complex hydrological applications such as cleanup of hazardous wastes, it is often necessary to remove some of the simplifying assumptions leading to Richards equation and return to the more complete multiphase model. Finally, it should be noted that in both the petroleum and groundwater remediation contexts, there may arise more complex situations in which mass transfer between phases and chemical reactions can occur. This creates the need for generalizations of the multiphase immiscible flow model presented here. However, that model serves as the basis for mathematically simulating fluid flow in a wide range of very important applications, and hence is used as the focus for the discussion here.

2.1 Single-Phase Flow

Much of the physical reasoning behind the immiscible multiphase model also underlies the single-phase flow model. Moreover, its presentation permits a context in which the important concept of absolute permeability can be highlighted.

The governing equations for the flow of a single compressible fluid through a porous medium (see, e.g., [15]) are the equation of continuity (reflecting the principle of conservation of mass) and Darcy's law:

$$\frac{\partial(\phi\rho)}{\partial t} = -\nabla \cdot (\rho\mathbf{v}) + \psi \tag{1}$$

$$\mathbf{v} = -\frac{k}{\mu}(\nabla p - \rho\mathbf{g}) \tag{2}$$

In this model, $p = p(\mathbf{x}, t)$ represents the pressure of the fluid at a point $\mathbf{x} \in \Omega \subset \mathbf{R}^3$ at time t. The density, ρ, and viscosity, μ, are fluid properties which are, in general, functions of the pressure. The porosity of the medium is represented by ϕ; it is the fraction of void space (e.g., the pore space, through which the fluid can flow) to total volume. The term $\mathbf{g}$ represents the gravitational acceleration vector, and ψ is a source/sink term.

The absolute permeability of the medium is represented by k (in general a second order tensor), and is defined through Darcy's law (2). This law is based on the empirical observation that the volumetric flow rate per cross-sectional area in the direction normal to the flow is directly proportional to the pressure gradient and inversely proportional to the viscosity. This relationship also defines the Darcy vector $\mathbf{v}$, which is the macroscopic fluid velocity.

As they are written, equations (1) and (2) express density, as depending on the velocity, and the velocity, as depending on the pressure gradient. In order for the system to be closed, it is necessary that a relationship between pressure and density be specified. This relationship, the thermodynamic equation of state, relates temperature, pressure and density.

2.2 Multiphase Flow

The general model used to describe flow of up to three immiscible fluid phases is now considered. Each of these phases has its own properties and quantities such as viscosity, density, etc. The fluid phases will be designated as wetting (w), nonwetting (nw), and intermediate wetting (iw). This designation is usually chosen according to the relative affinities of the fluids for the porous medium, but it can be chosen arbitrarily. The subscript i will be used generically, to stand for any of w, nw or iw.

In a given differential volume element, the fraction of the total amount of fluid present which consists of phase i is called the saturation of phase i, and is denoted S_i. From this definition we have

$$\sum_{i=1}^{3} S_i = 1. \tag{3}$$

Note that by this identity, among the three saturation variables, there are only two degrees of freedom.

The ith fluid phase has its own Darcy vector which, analogously to (2), is given by

$$\mathbf{v}_i = -\frac{k_i}{\mu_i}\left(\nabla p_i - \rho_i \mathbf{g}\right). \tag{4}$$

The quantities k_i are the "effective" permeabilities. In contrast to the absolute permeability, these are not properties of the medium only. They reflect the fact that the presence of each fluid phase interferes with the flow of the other. This effect is isolated mathematically by the introduction of the idea of relative permeability k_{ri}, defined by

$$k\,k_{ri} = k_i. \tag{5}$$

The relative permeablities k_{ri} are scalar functions of the fluid saturations, and typically do not exceed unity. They can depend on position as well, but are generally expected to exhibit much less spatial variability than the absolute permeability.

In addition to Darcy's law, each of the fluid phases must satisfy the principle of conservation of mass, which for each phase leads to a continuity equation analogous to (1). With $\mathbf{v}$ from (4) inserted into the continuity equation, the multiphase flow equations can be written as the following system of nonlinear evolution equations: for $i = w, nw, iw$,

$$\frac{\partial(\phi \rho_i S_i)}{\partial t} = \nabla \cdot \left[\,\frac{\rho_i k k_{ri}}{\mu_i}(\nabla p_i - \rho_i \mathbf{g})\,\right] + \psi_{\mathbf{i}}. \tag{6}$$

As in the single-phase case, the pressure and density for each phase are related (along with temperature) by the thermodynamical equations of state. Thus for each phase, for the pair p_i and ρ_i there is only one unknown.

With the requirement (3) on the saturations and the equations of state for each phase, this leaves two degrees of freedom among the nine unknowns S_i, ρ_i and p_i, $i = w, nw, iw$. Thus in order to close the system it is necessary to specify two additional relationships. These relationships are the capillary pressure functions. They reflect the fact that two immiscibly mixed fluids will have different pressures due to surface tension. In general, the capillary pressure between fluids i and j, which is denoted here by π_{ij}, is the difference between the two fluid pressures. Thus, for example,

$$\pi_{nw,iw} = p_{nw} - p_{iw},$$
$$\pi_{nw,w} = p_{nw} - p_w,$$

and $\pi_{iw,w}$ can be expressed in terms of these two. The capillary pressures are functions of the saturations. In principle, they can also vary spatially. Empirical observations relating capillary pressure to the absolute permeability [34] have been used to represent spatial variations in capillary pressure in numerical sentitivity studies [6].

These partial differential equations (1 and 2, or 6) are the basic equations used to describe the flow of up to three immiscible fluid phases in underground reservoirs. The state quantities are the pressures and saturations of each fluid phase, as a function of position and time. Boundary and initial conditions consistent with the physical problem under consideration must also be specified. Analytical solutions are available only under very restrictive conditions on geometry and fluid and rock properties, so numerical solutions are used. Due to the generally large scales associated with these types of problems, somewhat simplified representations may be considered for all or part of the studies. For example, one may assume that the flow can be suitably described with two, or perhaps a single, spatial dimension, or a relatively coarse numerical grid may be used. However, few realistic problems are suitably represented by a steady-state (time independent) formulation. While field-scale simulation is practiced routinely, the solution of these models remains an area of active research due to the desire for improvements in accuracy and efficiency of these large-scale problems ([12, 13, 16, 17, 18, 19]).

2.3 Estimation of Properties

In order for simulations with these models to be useful for predictive purposes, it is necessary that the physical properties of the media and fluids be specified with some degree of accuracy. Some of these are more readily determined than others. For example, properties depending only on the fluids, such as viscosity and the equations of state, can be determined prior to any consideration of porous media flow.

The porosity and absolute permeability are properties that depend on the rock, but do not depend on the fluids. These vary with position. Porosity is generally less variable than permeability, and can be measured fairly directly by well logging or experiments on reservoir core samples. On the other hand, the absolute permeability is actually defined through Darcy's law, and must be estimated through measures of pressure and flow rates observed during flow. Considerable attention has been directed to the estimation of the absolute permeability from data measured at reservoirs.

Traditionally, permeability estimates have been obtained from well-tests and reservoir history matching. Well-tests are essentially reservoir-scale experiments in which the fluid pressures at wells are observed in response to a perturbation in the flow rate. Historically, well-tests have been interpretted using simplified solutions, typically assuming uniform properties, and with graphically-based procedures [33]. Reservoir history matching involves the modification of properties within reservoir simulators so that predicted quantities, such as pressures at wells, match, in some sense, the observed historical performance. These type of data differ from well tests in that they are usually available over a longer period of time, and may include a number of wells. The flow encountered in these situations is usually much more complex than that in the controlled well tests. Reservoir history matching is a notoriously difficult problem, which has historically been carried out as a painstaking trial and error process. There have been considerable efforts at automation of this process (see, for example, [27, 32], and references therein). While approaches for analysis of well-tests and reservoir history have evolved along quite separate lines, they do represent quite similar inverse problems. It would be desirable to have firm theoretical bases and efficient methods for working these difficult problems.

In multiphase flow situations it is also necessary to specify the relative permeabilities and capillary pressure functions, which are properties of the fluids and the rock. These properties depend on the fluid saturations, and thus are functions of a state variable. Relatively little attention has been directed to the estimation of these properties from reservoir data (however, see [32, 52, 56, 58]). Most of the research regarding the estimation of properties from reservoir data has been directed to situations represented with the single-phase flow model. While this is a natural situation to study, it should be realized that many, if not most, of the situations for which reservoir simulations are desired are characterized by multiple fluid phases. In addition to the porosity and absolute permeability, the relative permeability and capillary pressure functions (or "multiphase flow functions") are unknown and subject to estimation in multiphase situations. It should also be noted that the sensitivity of quantities representing measured data to the porosity and absolute permeability will be different for the single and multiphase models, so that properties of the inverse problem associated with the single phase situations may not necessarily characterize the multiphase case [50].

The reliable estimation of relative permeability functions is difficult, even with experiments conducted on laboratory core samples. Although methods for estimating relative permeabilities from such experiments have been used for a number of years [25], serious problems associated with the interpretation of the experimental data have only more recently been exposed [41, 44]. These problems largely arise from the difficulty in eliminating various physical effects which were not included in the mathematical model upon which the analysis of experimental data was based. While considerable progress has been made in the methodology for interpretting such experiments [38, 55], recent experimental advances are providing exciting prospects for even more effective solutions on the laboratory scale. These advances are based on X-ray computed tomographic and magnetic resonance imaging methods [7, 11]. In transient, multiphase flow experiments with core samples, these imaging techniques can provide the saturation values for the different phases at thousands of points in the interior of the core sample at several consecutive times. This is extremely rich information, and can be used to identify the multiphase flow functions more accurately than has been previously possible [46].

3 Parameter Estimation Techniques

Since numerical methods must be used to solve the state equations, a finite-dimensional representation for the unknown properties suffices. For a selected parameterization of the unknown functions, a set of parameters are to be estimated from the measured data. We refer to this as the parameter estimation problem. Solution of the inverse problem, or estimation of the acutal properties, may involve solutions of multiple parameter estimation problems. In this section, the basic methods for solution of the parameter estimation problem are discussed.

3.1 Output Least Squares

In general, given a particular model, the parameter estimation or inverse problem is to find a parameter, denoted generically by q, which, when used in the model will cause the resulting "output" $F(q)$ of the model to match (approximately, at least) some observed data, denoted here by z_{OBS}. Thus the goal is to find q from within some admissible class Q_{AD} for which

$$F(q) \simeq z_{OBS}. \tag{7}$$

The range of the map F is denoted here by Z. Within this general notation, the output least squares approach is to seek to minimize the objective functional

$$J(q) = \frac{1}{2}\|F(q) - z_{OBS}\|_Z^2 \tag{8}$$

over Q_{AD}.

The data available for the inverse problems are invariably discrete, and are typically spatially sparse. For example, in a single-phase history matching problem, z_{OBS} might consist of observed pressures at one, or a few, wells within some time interval, $t \in [0, t_f]$. In multiphase situations, in addition to pressure data, temporal flow rates or cumulative productions of the individual fluid phases may also be available.

The selection of the admissible class Q_{AD} and the map $F(q)$ is an integral part of the estimation problem. For the single-phase estimation problem, one may consider the most general problem as one of estimating, say, a value of porosity and permeability corresponding to each grid block within the reservoir simulator. As stated, this problem would normally not be tractable as the number of unknowns would typically exceed the available data. On the other hand, conventional well-test methods are typically based on reservoir models whereby the porosity and permeability are represented as being homogeneous. The associated least squares problem would be relatively simple, involving just a few parameters, but the solution may not suitably represent the physical situation, and furthermore may not be useful for predictive purposes.

In the remainder of this section, we will discuss solution of the parameter estimation problem as defined by 8. We will then discuss further in the next section issues associated with the selection of the map $F(q)$.

3.2 Solution of Parameter Estimation Problem

There are two broad classes of schemes which have been used to solve the minimization problem. Methods in the first class are based on the Gauss-

Newton approach. Here, a sequence of iterates q_k is formed; at each q_k the parameter to state map $F(q)$ is linearized by

$$F(q_k + \delta q) \approx F(q_k) + F'_k \delta q, \tag{9}$$

where F'_k denotes the Fréchet derivative of $F(q)$ at q_k and is referred to as the "sensitivity coefficient" in the petroleum and hydrology literature. The map $F(q)$ in $J(q)$ is replaced by this linearization, resulting in a quadratic subproblem for δq. With the residual defined by $r_k \stackrel{\text{def}}{=} F(q_k) - z_{OBS}$, this may be written as

$$J(q_k + \delta q) \approx \tilde{J}_k(\delta q) \stackrel{\text{def}}{=} \frac{1}{2} \|F'_k \delta q + r_k\|^2_Z. \tag{10}$$

The next iterate q_{k+1} is then given by $q_{k+1} \stackrel{\text{def}}{=} q_k + \delta q_k$, where δq_k is a minimizer of (10). The first order necessary condition for (10) leads to

$$[(F'_k)^*(F'_k)]\delta q_k = -(F'_k)^* r_k. \tag{11}$$

This is the basic Gauss-Newton method. While it has the advantage of being quadratically convergent under certain conditions, in its unmodified form it is usually of little practical value due to lack of robustness. This is because typically the inverse problems under discussion here are ill-posed, often severely so. This ill-posedness manifests itself in the sensitivity coefficient F'_k in that the singular values of that operator typically cluster at zero. In fact it is often observed that they converge to zero at an exponential rate as the index goes to infinity. This was observed, e.g., in [47], where a linear model analogous to (1) and (2) was used in a simulated history matching problem. This property is a reasonable way of precisely defining "severely ill-posed".

An effective stategy for improving the robustness of this method is the use of trust region techniques [35]. These are based on the idea that one should only "trust" the quadratic model (10) within a certain radius β_k and so should constrain the linear subproblem accordingly. As implemented, this approach leads to the Levenberg-Marquardt method, one of the most popular methods for solution of nonlinear least squares problems. Comparisons between the basic Gauss-Newton method and the Levenberg-Marquardt method for reservoir history matching problems are discussed in [54], in which the superiority of the Levenberg-Marquardt method is clearly illustrated.

In numerical implementations of most Gauss-Newton type methods, a matrix representation of F'_k is explicitly formed. This is done column by column; each column represents a directional derivative of F. The simplest and perhaps most widely used technique for computing these directional derivatives is finite differencing:

$$F_k' \delta q \approx \frac{F(q_k + \tau \delta q) - F(q_k)}{\tau} \tag{12}$$

for some small $\tau > 0$. Thus each computation of a column of F_k' requires a separate solution of the state equations (1) or (6). For problems with relatively large degrees of freedom, efficient ways for computing F_k' are desired. The advent of parallelism in computer architecture enables this approach to be used with significantly larger degress of freedom. Since the computation of each directional derivative of F is independent of all other directional derivatives, they can be done in parallel in a very simple and natural way. Secondly, reduced subspace algorithms are being developed ([47, 48]) to solve (11) iteratively rather than directly. These do not require the explicit formation and storage of F_k', only the ability to apply it and its adjoint to a given vector. The adjoint computation is discussed below.

Methods in the second main class for solution of the minimization problem employ the gradient, but do not directly utilize F_k'. These methods utilize an adjoint method for efficient calculation of the gradient vector. This approach was introduced as an application of ideas from optimal control theory [9, 10]. In that point of view, the dependence of the least-squares problem on the model is not viewed in terms of what is here called the map F, but rather as a constraint on the optimization problem. The constraint is the dynamical system governing the model, or state equations. This point of view leads to the use of Lagrange multiplers and the Pontryagin Maximum Principle, the cornerstone of optimal control theory. In that context, the gradient calculation is given in terms of the solution of a certain linear dynamical system, known as the costate equation, occuring in the dual space of the space in which the state equation is posed. For these reasons, the adjoint approach to computing the gradient of $J(q)$ is often referred to as the costate or optimal control approach.

The point of this approach is that, given a $q \in Q_{AD}$ and the corresponding solution of the state equation, the bulk of the effort in computing $\nabla J(q)$ lies in the solution of the costate equation. This can be dramatically less costly than the explicit computation of F_k'. Disadvantages of this approach are that it requires development and solution of a system comparable in size to the forward problem, and that the minimization is limited to methods which may be somewhat less efficient (in terms of the number of iterations for solution) and less robust than the Levenberg-Marquardt method. Quasi-Newton and some conjugate gradient algorithms [30, 57] appear to be good choices for use with the adjoint approach. A number of other applications of the costate approach have appeared in the petroleum and hydrology literature (see, e.g., [4, 5, 31, 32, 39, 40, 43, 49, 56, 58]).

4 Ill-Conditioning and Proposed Remedies

In Section 3.2, the ill-conditioning of the parameter estimation problem was mentioned in the context of the solution of the linear subproblems (11) and the lack of robustness of the basic Gauss-Newton method. There the Levenberg-Marquardt method was briefly discussed as a popular and reliable modification of that method. However, the difficulty is more fundamental: if the linear subproblems are ill-posed it is generally due to the ill-posedness of the original inverse problem.

The minimization problem (8) is said to be well-posed in the sense of Hadamard if the solution *(i)* exists, *(ii)* is unique, and *(iii)* depends stably (continously) on z_{OBS}. If any of these conditions fail, then the problem is ill-posed.

Any of these three conditions can easily fail if the inverse problems are not formulated carefully. This is particularly true of field-scale history matching problems. For example, consider the estimation of the absolute permeability $k(x)$ on the basis of pressure history $p(x_0, t)$ at a well. (Thus q and z_{OBS} in the general setting are $k(\cdot)$ and $p(x_0, \cdot)$, respectively.) It is clearly not possible to recover an arbitrary $k(x)$ over a two- or three-dimensional spatial domain, from the one-dimensional measurement $p(x_0, t)$; clearly non-uniqueness is possible. On the other hand, due to noise in the data, it can happen that J has no minimizer at all. Finally, the stability condition (iii) can easily fail because two very different $k(\cdot)$ can map to two $p(x_0, \cdot)$ which are "close", so that the inversion is unstable.

Various strategies can be employed to formulate the estimation problem so that ill-posedness can be mitigated or eliminated. These are based on these essential ideas: (1) restrict or reduce the parameter space of the unknowns, or (2) incorporate prior knowledge regarding parameters. The first idea can be implemented through selection of parameterizations of the unknown functions, or by augmentation of the objective functional with regularization terms. The second idea is also implemented with additional terms within the objective functional in a way that is computationally similar to regularization. In §4.1, we present strategies that retain (8) as the objective functional. In §4.2, we consider methods that lead to other objective functionals.

4.1 Restriction of Parameter Space

Many inverse problems of practical interest can be posed as maximum likelihood problems through selection of the norm associated with (8) [53]. Statistical principles can then be used to guide selection of parameterizations of the unknown functions in a way that is consistent with the measured data. This can be approached as a system identification problem–that is, determination of the appropriate model, as well as the parameters within the model, that satisfies the measured data.

Watson et al. [53] have presented a procedure to select among candidate models. The candidate models correspond to different parameterizations of the unknown functions used within the general equations discussed in §2. The strategy is based on the principle of parsimony [2]: seek the simplest possible model, in terms of the fewest degrees of freedom, which fully satisfies the data, and is consistent with all other available knowledge of the system. In analyzing well-test and reservoir production data, they formed a hierarchy of candidate models, or parameterizations, consistent with other geological or engineering information about the reservoir. Many of the models were nested, in that simpler models could be obtained by selecting certain values for parameters in more complex models. Statistical hypotheses were evaluated with the F-test to ascertain the significance of the additional paramters in the more complex of the nested models. If the additional parameters were not statistically significant, the more complex model was discarded as a candidate. The ability of the selected model to suitably describe the data can be evaluated through residual analyses [53].

This procedure may not always lead to selection of a single candidate, and it can be tedious if a relatively large number of degrees of freedom are required to adequately resolve the data. However, it does provide a sound method for substantially reducing the number of candidate solutions to the inverse problem, and it enables incorporation of prior information regarding the properties that may be available from a variety of sources. The method has been used to design well-tests for detecting certain properties by evaluating conditions for identifiability, and it can also be used as a guide to selection of zonation, a standard approach used by reservoir engineers to enforce the degree of smoothness of the reservoir properties.

An application of this procedure to well-test data is shown in Figure 1. The data consisted of down-hole pressures measured over time during a pressure-buildup test. The full set of data were analyzed, but the example here is based on analysis of just the first 36 data points. The conventional method for analyzing such data, which is graphically based, would lead to selection of a reservoir model represented by uniform values of porosity and permeability. Since the formation may be naturally fractured, a dual-porosity reservoir model [51], which represents the media as being composed of two regions with different properties, was also considered as a candidate. Both the single-porosity and dual-porosity models provided fairly precise fits of the data, but the F-test analysis indicated that the additional parameters in the dual-porosity reservoir model were significant. Predictions of the future values of pressures using estimates of the single and dual-porosity models showed that the selected dual-porosity model was a much better predictor of reservoir behavior (see Figure 1). This example illustrates the importance of model selection for prediction of future reservoir behavior.

A particularly challenging problem has been the estimation of the functions which arise in the description of multiphase flow–the relative permeability and capillary pressure functions. These properties are somewhat unique in the field of estimation since they represent functions of the state, or dependent variables, of the model. A procedure has been developed for estimating these functions within the context of estimating relative permeability and capillary pressure functions from laboratory experiments [38, 55]. The unknown functions are represented by B-splines, due to their ability to accurately represent any smooth functions and their computational convenience. A series of estimation problems are solved, in which the spline dimensions are increased. The selected representation corresponds to that with the fewest degrees of freedom for which essentially the smallest residual value of the objective function is attained. An important consideration is that sufficient degrees of freedom are provided so that bias errors associated with selection of the functional representations can be avoided [26]. Once the data are satisfied, in that the mismatch of the data by predicted quantities can be attributed to random errors, further increase in the degrees of freedom would only serve to increase the uncertainty associated with the estimates.

This procedure is illustrated for analysis of experimental data collected during a laboratory experiment on a core sample [55]. The data consisted of values of the pressure drop and production of the displaced fluid measured while the initial saturating fluid phase is displaced with a second fluid phase. A series of parameter estimation problems were solved as the dimensions of the splines representing the unknown relative permeability curves were increased. A plot of the residual objective function value as a function of the degrees of freedom showed a characteristic sharp decline, followed by a leveling of the curve. Use of the procedure discussed in the previous paragraph led to selection of splines with a total of twelve parameters. These estimates are shown in Figure

2 along with estimates corresponding to eight and ten parameters, and those obtained with a power-law model (two parameters). This figure illustrates that the power-law model provides poor estimates of the unknown functions. It also shows that selection of the specific numbers of degrees of freedom for this problem is not critical, provided sufficient degees of freedom have been provided. That is, the bias error can be largely eliminated, while the experimental design provides sufficient information to limit the variance error [26]. This approach has been extended to a variety of problems in which two and three-phase flow functions are to be estimated from experimental data [36, 38, 46].

4.2 Regularization and Bayesian Estimation

Other approaches may be more appropriate when dealing with the truly large scale estimation problems for which suitable candidate parameterizations of the spatially variable funcitons may not be obvious, or solutions of large numbers of parameter estimation problems are to be avoided. Two basic approaches, regularization and Bayesian estimation, have been used. The methods in this section differ from those discussed previously in that there have as yet been relatively few applications with actual reservoir or laboratory data.

The regularization approach is based on ideas first advanced by Tikhonov [45] in the solution of ill-posed integral equations. As implemented by Seinfeld and coworkers [28, 29, 30], for some $\beta \geq 0$ the objective functional is augmented with a regularizing functional $\| \cdot \|_R^2$ which can be based on a Sobolev seminorm or can incorporate statistical beliefs as discussed below:

$$J_\beta(q) = J(q) + \frac{\beta}{2}\|q\|_R^2. \tag{13}$$

Implementations have been demonstrated for the estimation of the porosity and permeability for hypothetical single- and two-phase reservoir estimation problems [30, 31, 32]. In these studies, the unknown functions were represented on a two-dimensional grid with bicubic splines. Important considerations in this approach are the selection of the norm for the stabilizing term, the weighting of the stabilizing term β, and the grids for the representation of the unknown properties [31]. The weighting of the stabilizing term can be particularly important. Reported methods (see, e.g., [24, 31, 58]) require multiple solutions of parameter estimation problems.

Frequently in these reservoir estimation problems there are other data in addition to the flow data measured at the wells that may be available. Such information can include data collected from laboratory experiments on reservoir core samples, property estimates from logging or well-test data, and general information from geological studies. It is most desirable that the estimates honor such prior knowledge of the system, when appropriate, or that this knowledge be incorporated into the estimation problem in order to improve the conditioning. This has led to Bayesian-based approaches for estimating the unknown parameters [3, 23, 58, 59].

Prior knowledge of the system can be incorporated into the problem by way of a modified regularization. In particular, given a q_0 which embodies prior information or beliefs about the solution, the regularized function J_β can be changed to

$$J_\beta(q) = J(q) + \frac{\beta}{2}\|q - q_0\|_R^2. \tag{14}$$

This approach amounts to a Bayesian estimation under certain conditions. In particular, suppose the data z_{OBS} are related to a $\bar{q}$ by

$$z_{OBS} = F(\bar{q}) + \varepsilon_z$$

and that q_0 satisfies

$$\bar{q} = q_0 + \varepsilon_q$$

for random variables ε_z and ε_q. If the underlying probability distributions on these random variables are Gaussian, then minimization of J_β as given by (14) has a Bayesian interpretation for suitably chosen $\|\cdot\|_z$, $\|\cdot\|_R$, and β. In particular, if ε_z and ε_q have mean zero and convenience matrices $\sigma^2 V$ and $\tau^2 \Sigma$, respectively, and if

$$\begin{aligned}
\|z\|_z^2 &= z^T V^{-1} z, \\
\|q\|_R^2 &= q^T \Sigma^{-1} q, \\
\beta &= \sigma^2/\tau^2,
\end{aligned}$$

then minimization of J_β as given in (14) is a Bayesian maximum likelihood estimation. This expression for β is especially enlightening: it shows that the optimal β is that which equally balances the weight given to the mean-squared error in the z_{OBS} and the mean squared error in the q_0.

See [22] for an excellent discussion of the relationship of regularization to Bayesian estimation and [3, 5, 23, 27, 58, 59] for applications of these ideas in petroleum and hydrology.

We conclude here with an example that serves to illustrate the desirability of using prior information in two-phase reservoir diplacement situations. Consider the displacement of oil by water. It is desired that the reservoir parameters be estimated relatively early in the displacement process so that reliable predictions of reservoir performance can be made. Early in the displacement process, the fluid saturations of the displacing fluid will be relatively small, so more information would be reflected about the relative permeability curves corresponding to lower saturation ranges. However, predictions are desired for situations with relatively larger saturations. If prior estimates of these functions are available, they can be included in the estimation problem through the Bayesian term. If the relative weighting of that term is appropriately chosen, one would hope that regions of the function that are relatively well determined be estimated through the measured reservoir data, while those regions that are not well determined should not deviate significantly from their prior estimates. A hypothetical test problem shows that this can be accomplished [58]. Estimates for the relative permeability curves determined by minimizing the least-squares functional are shown in Figure 3. They indicate that the water relative permeability corresponding to the larger saturation range are poorly determined. When the estimation is performed including prior estimates, that region corresponds closely to the prior estimates, whereas the well-determined region is closer to the true values (see Figure 4). Figure 5 shows that predictions with the Bayesian estimates are considerably better than those obtained with the prior estimates. It is also interesting to note how sensitive the solution is to the shape of the relative permeability curves. Although the true curves and prior estimates do not differ so significantly in value at any given saturation (see Figure 5), the simulated pressures differ quite significantly. The method for estimating the relative weighting of the Bayesian term was accomplished through solution of several minimization problems solved using the BFGS algorithm. Since each of the problems was relatively well conditioned, and information regarding the Hessian matrix was built up through the process, the entire problem was solved more efficiently than a single minimization of the least-squares objective functional.

5 Conclusions

The estimation of reservoir properties is very important and challenging inverse problem. Effective solutions to this problem require consideration of the ill-posedness inherent to the problem. Three basic approaches were identified. The use of statistical concepts to guide selection of parameterizations of the unknown functions provided effective solutions for problems which could be suitably described with relatively few degrees of freedom. Regularization and Bayesian approaches have been advanced for the truly large-scale reservoir history matching problems. Further research into these problems is desirable in order to foster application of systematic methods for actual large-scale problems.

Acknowledgments

One of the authors (ATW) thanks Jan-Erik Nordtvedt for helpful comments during preparation of this article

References

[1] J. Bear, *Dynamics of Fluids in Porous Media*, Elsevier, 1972.

[2] G.E.P. Box and G.M. Jenkins, *Time Series Analysis: Forecasting and Control*, Holden-Day Inc., 1970.

[3] J. Cararra and S.P. Neuman, Estimation of aquifer parameters under transient and steady state conditions: 1. Maximum likelihood method incorporating prior information, *Water Resources Research* **22(2)** (1986), 199–210.

[4] J. Cararra and S.P. Neuman, Estimation of aquifer parameters under transient and steady state conditions: 2. Uniqueness, stability and solution algorithms, *Water Resources Research* **22(2)** (1986), 211–227.

[5] J. Cararra and S.P. Neuman, Estimation of aquifer parameters under transient and steady state conditions: 3. Application to synthetic and field data, *Water Resources Research* **22(2)** (1986), 228–242.

[6] J. Chang and Y.C. Yortsos, Effect of capillary heterogeneity on Buckley-Leverett displacement, *SPE Reservoir Engineering*, May 1992, 285–293.

[7] C. Chardaire-Rivière, G. Chavent, J. Jaffré, J. Liu, and B.J. Bourbiaux, Simulataneous estimation of relative permeabilities and capillary pressure, *SPE Formation Evaluation*, December 1992, 283–289.

[8] G. Chavent, Identification of distributed parameter systems: About the output least squares method, its implementation and identifiability, *Identification and System Parameter Estimation, Proceedings 5th IFAC Symp., Darmstadt, FRG)* (R. Isermann, ed.), Pergamon, 85–97.

[9] G. Chavent, M. Dupuy, and P. Lemonnier, History matching by use of optimal control theory, *SPE Journal*, February, 1975, 74–86.

[10] S. Chen, F. Qin, K.-H. Kim, and A.T. Watson, NMR imaging of multiphase flow in porous media, *AIChE Journal* **39** (1993), 925–934.

[11] W.H. Chen, G.R. Gavalas, J.H. Seinfeld, and M.L. Wasserman, A new algorithm for automatic history matching, *SPE Journal*, December 1974, 593–608.

[12] H.K. Dahle, M.S. Espedal, R.E. Ewing, and O. Sævareid, Characteristic adaptive sub-domain methods for reservoir flow problems, *Numerical Methods for Partial Differential Equations* **6** (1990), 279–309.

[13] M. Espedal and R.E. Ewing, Characteristic Petrov-Galerkin subdomain methods for two-phase immiscible flow, *Computer Methods in Applied Mechanics and Engineering* **64** (1987), 113–135.

[14] R.E. Ewing, Determination of coefficients in reservoir simulation, in *Numerical Treatment of Inverse Problems for Differential and Integral Equations* (P. Deuflhardt and E. Hairer, eds.), Birkhauser, Berlin, 1982, 206–226.

[15] R.E. Ewing, Problems arising in the modeling of processes for hydrocarbon recovery, in *The Mathematics of Reservoir Simulation* (R.E. Ewing, ed.), *SIAM Frontiers in Applied Mathematics* **1**, SIAM, Philadelphia, PA, 1983.

[16] R.E. Ewing, Finite element methods for nonlinear flows in porous media, *Computer Meth. Appl. Mech. Eng.* **51** (1985), 421–439.

[17] R.E.Ewing, Operator splitting and Eulerian-Lagrangian localized adjoint methods for multiphase flow, *The Mathematics of Finite Elements and Applications VII MAFELAP 1990* (J. Whiteman, ed.), Academic Press Inc., San Diego, California, 1991, 215–232.

[18] R.E. Ewing, Finite element methods for multiphase and multicomponent flows, *Finite Elements in Fluids* **8** (T.J. Chung, ed.), Hemisphere Publishing Corporation, Washington, DC, 1992, 165–176.

[19] R.E. Ewing, B.A. Boyett, D.K. Babu, and R.F. Heinemann, Efficient use of locally refined grids for multiphase reservoir simulation, *SPE 18413, Proceedings Tenth SPE Symposium on Reservoir Simulation*, Houston, Texas, February 6–8, 1989, 55–70.

[20] R.E. Ewing and J.H. George, Identification and control for distributed parameters in porous media flow, *Distributed Parameter Systems, Lecture Notes in Control and Information Sciences* **75** (M. Thomas, ed.), Springer-Verlag, May 1985, 145–161.

[21] R.E. Ewing and T. Lin, A class of parameter estimation techniques for fluid flow inporous media, *Advances in Water Resources* **14(2)** (1991), 89–97.

[22] B.G. Fitzpatrick, B.G., Bayesian analysis in inverse problems, *Inverse Problems* **7** (1991), 675–702.

[23] G.R. Gavalas, P.C. Shah, and J.H. Seinfeld, Reservoir history matchng by Bayesian estimation, *SPE Journal* **16** (1976), 337–350.

[24] P.C. Hansen, Analysis of discrete ill-posed problems by means of the L-curve, *SIAM Review* **34** (1992), 561–580.

[25] E.F. Johnson, D.P. Bossler, and V.O. Naumann, Calculation of relative permeability from displacement experiments, *Trans., AIME* **216** (1959) 61–63.

[26] P.D. Kerig and A.T. Watson, Relative permeability estimation from displacement experiments: An error analysis, *SPE Reservoir Engineering* **1** (1986), 175–182.

[27] J.B. Kool, J.C. Parker, and M.T. Van Genuchten, M.T., Parameter estimation for unsaturated flow and transport models — a review, *Journal of Hydrology* **91** (1987), 255–293.

[28] C. Kravaris and J.H. Seinfeld, J.H., Identification of parameters in distributed parameter systems by regularization, *SIAM J. Control. Optim.* **23** (1985), 217–241.

[29] C. Kravaris and J.H. Seinfeld, Identification of spatially varying parameters in distributed paramter systems by discrete regularization, *J. Math. Anal. Appl.* **119** (1986), 128–152.

[30] T. Lee, C. Kravaris, and J.H. Seinfeld, History matching by spline approximation and regularization in single-phase areal reservoirs, *SPE Reservoir Engineering*, September 1986, 521–534.

[31] T. Lee and J.H. Seinfeld, Estimation of two-phase petroleum reservoir properties by regularization, *J. of Comp. Phy.* **69** (1987), 397–419.

[32] T. Lee and J.H. Seinfeld, Estimation of absolute and relative permeabilities in petroleum reservoirs, *Inv. Prob.* **3** (1987), 711–728.

[33] W.J. Lee, *Well Testing*, SPE, Richardson, TX, 1982.

[34] M.C. Leverett, Capillary Behavior in Porous Solids, *Trans. AIME* (1941), 152–167.

[35] J.J. More, The Levenberg-Marquardt algorithm: Implementation and theory, *Numerical Analysis, Proceedings, Biennial Conference Dundes* (G.A. Watson, ed.), Springer-Verlag, 1977.

[36] J.E. Nordtvedt, G. Mejia, P. Yang, and A.T. Watson, Estimation of capillary pressure and relative permeability from centrifuge experiments, *SPE Reservoir Engineering*, (to appear).

[37] O.A. Plumb and Whitaker, Diffusion, adsorption and dispersion in porous media: Small-scale averaging and local volume averaging, in *Dynamics of Fluids in Hierarchical Porous Media* (J.H. Cushman, ed.), Academic Press, 1990.

[38] P.C. Richmond and A.T. Watson, Estimation of multiphase flow functions from displacement experiments, *SPE Reservoir Engineering* **5** (1990), 121–127.

[39] J.H. Seinfeld and C. Kravaris, Distributed parameter identifiction in geophysics – petroleum reservoirs and aquifers, in *Distributed Parameter and Control Systems* (S.G. Tzafestas, ed.), Pergamon, 1982.

[40] P.C. Shah, G.R. Gavalas, and J.H. Seinfeld, Error analysis in history matching: optimum level of parametrization, *SPE Journal* **18** (1978), 219–228.

[41] P.M. Sigmund and F.G. McCaffery, An improved unsteady-state procedure for determining the relative-permeability characteristics of heterogeneous porous media, *SPE Journal*, February 1991, 15–28.

[42] J.C. Slattery, *Momentum, Energy, and Mass Transfer in Continua*, Kreiger, 1981.

[43] N.-Z. Sun and W.W.-G. Yeh, Coupled inverse problem in groundwater modeling, 1. Sensitivity analysis and parameter identification, *Water Resources Research* **26** (1990), 2507–2525.

[44] T.M. Tao and A.T. Watson, Accuracy of JBN estimates of relative permeability, Part 1: Error analysis, *SPE Journal* **24** (1984), 215–223.

[45] A.N. Tikhonov and V.Y. Arsenin, *Solutions of Ill-Posed Problems*, Wiley, 1977.

[46] G.M. Valazquez, *A Method for Estimating Three-Phase Flow Functions*, Ph.D. Dissertation, Texas A&M University, College Station, Texas, May 1992.

[47] C.R. Vogel and J.G. Wade, A modified Levenberg-Marquardt algorithm for large-scale inverse problems, *Proceedings of the Conference on Computation and Control III*, Bozeman, Montana, August 1992.

[48] C.R. Vogel and J.G. Wade, Iterative SVD-based methods for ill-posed problems, *SIAM Journal on Scientific and Statistical Computing*, (to appear).

[49] M.L. Wasserman, A.S. Emanuel, and J.H. Seinfeld, Practical applications of optimal control theory to history matching multiphase simulator models, *SPE Journal*, August 1975, 347–355.

[50] A.T. Watson, Sensitivity analysis of two-phase reservoir history matching, *SPE Reservoir Engineering* **4** (1989), 319–324.

[51] A.T. Watson, J.M. Gatens III, W.J. Lee, and Z. Rahim, An analytical model for history matching naturally fractured reservoir production data, *SPE Reservoir Engineering* **5** (1990), 384–388.

[52] A.T. Watson, G.R. Gavalas, and J.H. Seinfeld, Identifiability of estimates of two-phase reservoir properties in history matching, *SPE Journal* **64** (December 1984), 697–706.

[53] A.T. Watson, H.S. Lane,and J.M. Gatens III, History matching with cumulative production data, *Journal of Petrolem Technology* **42** (1990), 96–100.

[54] A.T. Watson and W.J. Lee, A new algorithm for automatic history matching production data, *SPE Unconventional Gas Technology Symposium*, 1986, 235–244.

[55] A.T. Watson, P.C. Richmond, P.D. Kerig, and T.M. Tao, A regression-based method for estimating relative permeabilities from displacement experiments, *SPE Reservoir Engineering* **3** (1988) 953–958.

[56] A.T. Watson, J.H. Seinfeld, G.R. Gavalas, and P.T. Woo, History matching in two-phase petroleum reservoirs, *SPE J.*, December 1980, 521–532.

[57] P.-H. Yang and A.T. Watson, Automatic history matching with variable metric methods, *SPE Reservoir Engineering*, August 1988, 995–1001.

[58] P.-H. Yang and A.T. Watson, A Bayesian methodology for estimating relative permeability curves, *SPE Reservoir Engineering*, May 1991, 259–265.

[59] W. Yeh, Y.S. Moon, and K.S. Lee, Aquifer parameter identifiction with kriging and optimum parameterization, *Water Resources Reseach* **19(1)** (1983), 225–233.

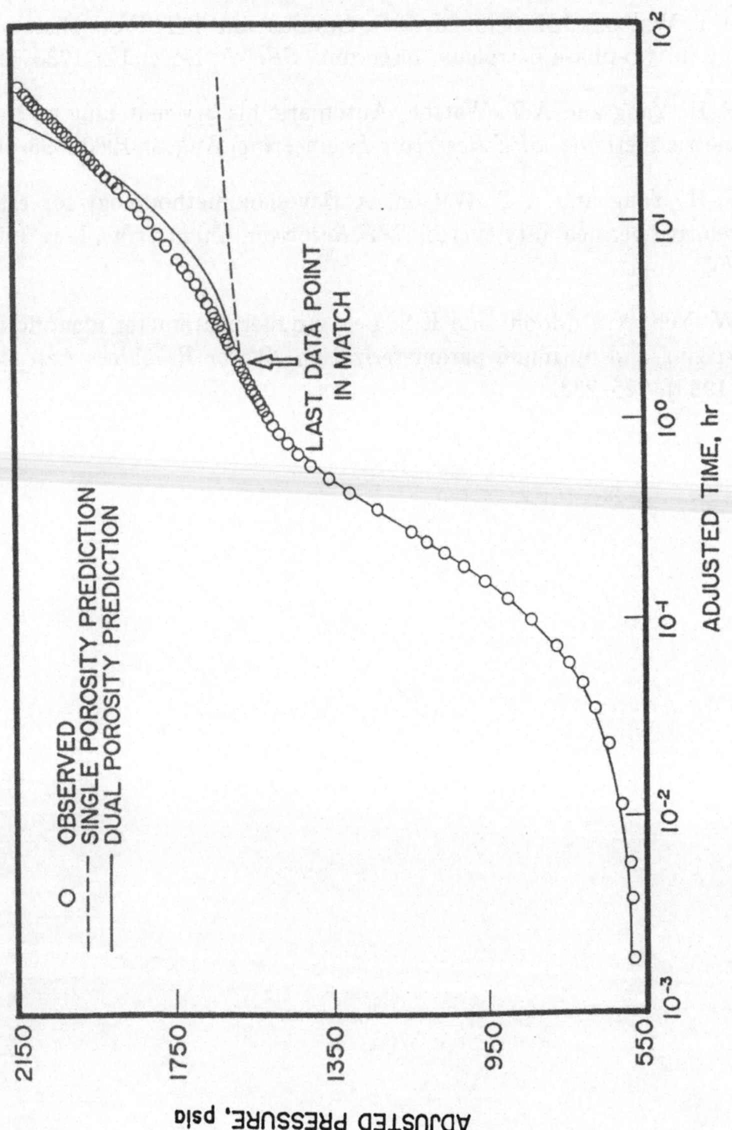

Figure 1. Measured well-test pressure data with calculated and predicted quantities. Copyright SPE. Reprinted with permission [53].

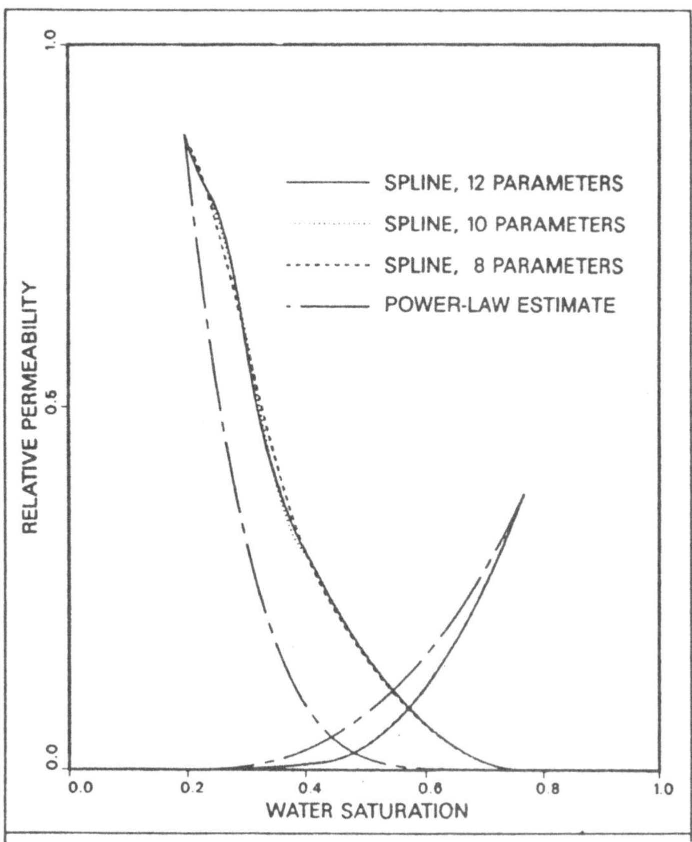

Figure 2. Estimates of relative permeability functions from laboratory experimental data. Copyright SPE. Reprinted with permission [55].

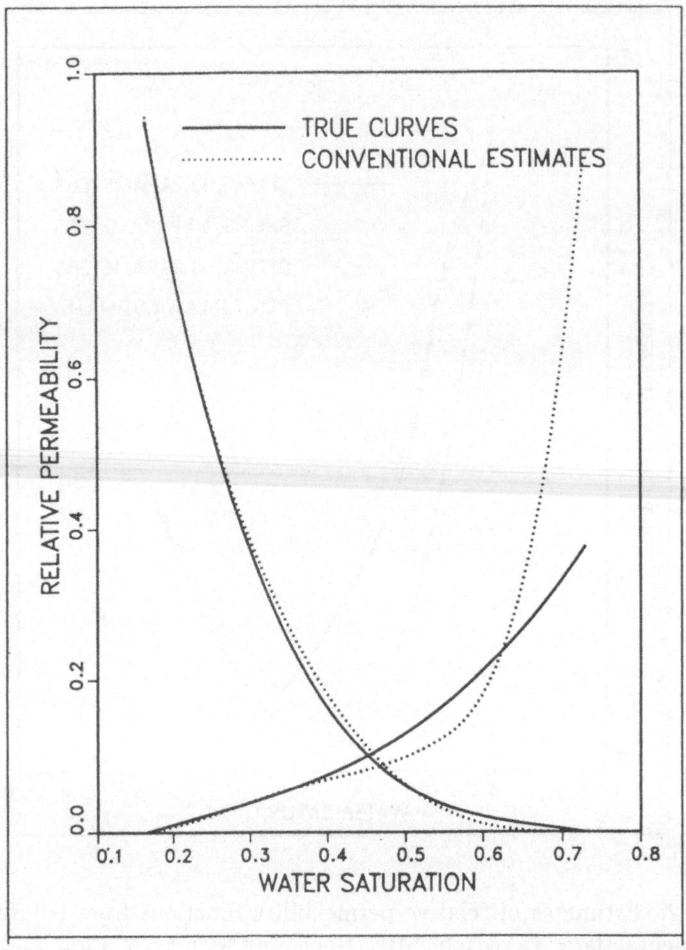

Figure 3. Least squares estimates of relative permeability functions for hypothetical waterflood. Copyright SPE. Reprinted with permission [58].

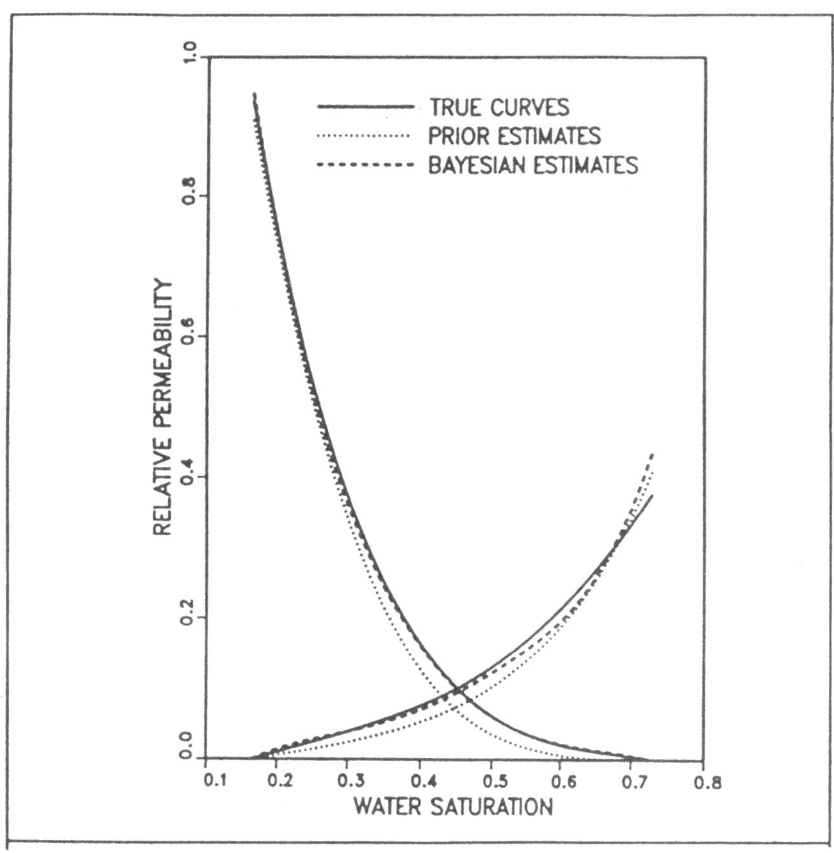

Figure 4. Bayesian estimates of relative permeability functions for hypothetical waterflood. Copyright SPE. Reprinted with permission [58].

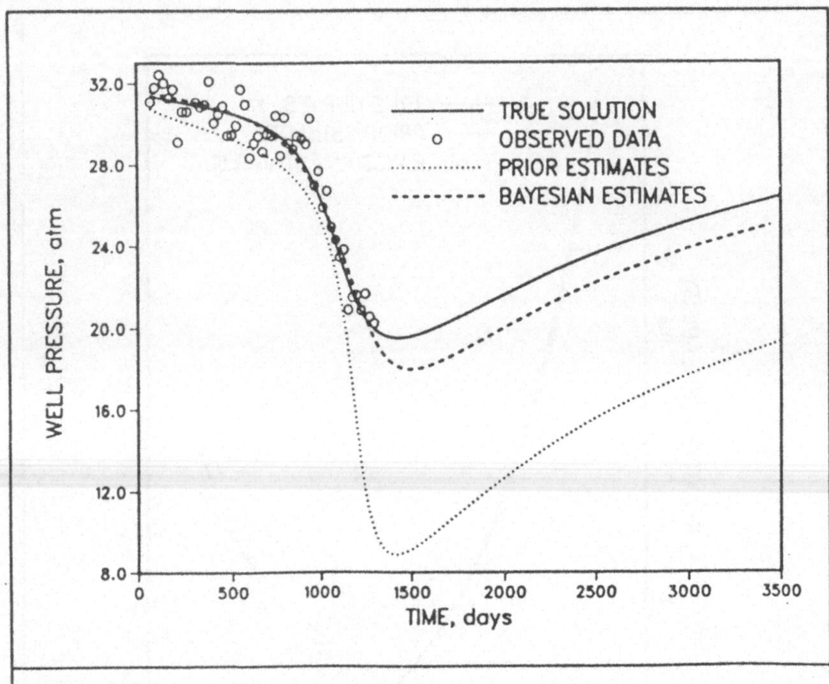

Figure 5. Measured pressure data with calculated and predicted quantities for hypothetical waterflood. Copyright SPE. Reprinted with permission [57].

DETERMINATION OF PETROPHYSICAL PARAMETERS BY RESOLVING AN INVERSE PROBLEM

Catherine Rivière - Institut Français du Pétrole - France

1 - INTRODUCTION

In the petroleum industry, when a project is drawn up to make a field produce or during the monitoring of the application of this project, it is indispensable to simulate fluid flow inside the field. For this, the laws governing multiphase flows are not always satisfactory, particularly when all three phases are mobile (three-phase flow).

However, the laws governing two-phase flows are quite well known, and they can be used to model a great many phenomena. These laws are also used for the laboratory-scale modeling of fluid flow to understand the phenomena involved and to then be able to scale them up to field scale.

Important parameters used in simulations are petrophysical parameters called relative permeabilities and capillary pressure. There are functions of saturation.

These parameters can be measured directly by long and expensive experiments or can be determined by adjustment methods.

This paper describes an automatic method for simultaneously estimating both relative permeabilities and capillary pressure from a single experiment.

Simulation of multiphase flow in porous media requires knowledge of relative permeabilities. Such data can be obtained by laboratory experiments. Some methods are time-consuming, and several days are needed to acquire a complete set of relative permeabilities, others methods are limited to displacements where the capillary pressure must be ignored. Another limitation of these methods is that relative permeabilities are not calculated in the full saturation range. To improve these methods, analytical and adjustment techniques have been developed to take capillary pressure into account and to obtain relative permeabilities for the overall saturation range.

Adjustment methods infer relative permeabilities from pressure drop and production data obtained during laboratory experiments. Different simulations of a reservoir model are compared to calculate a single set of relative permeabilities by history matching of production and pressure data. This can be done automatically with a nonlinear least-squares approach to match the pressure drop and production history. The different automatic adjustment methods currently available may or may not account for capillary pressure and require one or more experiments.

Chavent *et al.*[1] proposed an automatic adjustment method. Two experiments are needed for the interpretation: one with a high flow rate, which allows capillary pressure to be ignored, and one with a low flow rate to take capillary effects into account. The experimental data required are the pressure drop and the cumulative volume of fluid produced as functions of time.

Kerig and Watson[2] estimated relative permeabilities from one displacement experiment. The data measured are the pressure drop across the core sample and the cumulative volume of displaced phase recovered. The relative permeabilities are represented by cubic spline. Yang and Watson[3] again used the previous method, adding a Bayesian-type performance index to incorporate prior estimates of relative permeability curves into an automatic history-matching algorithm. The method was tested on hypothetical waterfloods, and good results were obtained. In Ref. 3, the capillary pressure was ignored, but the same method also was used to estimate relative permeability and capillary pressure curves simultaneously[4].

Capillary pressure curves usually are obtained from laboratory tests with various techniques. We describe in this paper a method for the simultaneous estimation of relative permeabilities and capillary pressure. This method incorpoates the following new features with respect to previous work[1].

1. It is now possible to measure saturation profiles experimentally along the core sample by gamma ray attenuation technique or computerized tomography scanning in addition to the pressure drop and the production data.

2. A new scheme with slope limiters and implicit discretization for capillary diffusion has been developed[5,6]. This scheme has low numerical dispersion and good stability and closely approaches the entropic solution.

3. Development time and reliability of the code are improved by use a symbolic program to generate the corresponding FORTRAN code from the input of the discretized aquations and for the error function[7].

As in previous work[1,8] the optimal control theory is used to solve the minimization problem in the least-squares method. Compared with Richmond and Watson's[4] approach, our method uses additional measurements of saturation profiles. These additional data make the optimization problem easier to solve, especially for the estimation of capillary pressure.

The basic models used to describe the different laboratory experiments studied are discussed, and them the parameter estimation method. Finally, different test cases are presented.

2 - BASIC MODELS

Two-phase displacements in laboratory experiments are simulated with a 1D, two-phase incompressible model based on Darcy's law, together with various boundary conditions enabling the simulation of various laboratory experiments.

The main equation is:

$$\Phi \frac{\partial S}{\partial t} - \frac{\partial}{\partial x} Ka(S) \frac{\partial S}{\partial x} + q_T(t) \frac{\partial}{\partial x} \left(\frac{ko}{ko + k\omega} \right)(S)$$

$$+ \frac{\partial q_G}{\partial x} b_G(S) = 0 \tag{2.1}$$

$$q_T = -K(k\omega + ko)(\nabla p - \frac{ko\rho o + k\omega\rho\omega}{2(\rho o + \rho\omega)} g \nabla z \tag{2.2}$$

with q_T : total flow rate independent of x

$$q_G = K(\rho o + \rho\omega)g \frac{\nabla z}{2}: \text{ gravity field} \tag{2.3}$$

where S is the normalized water saturation and P the global pressure[5] .

with :

$$b_G = \frac{ko\,k\omega}{ko + k\omega} \frac{\rho\omega - \rho o}{2(\rho\omega - \rho o)} \tag{2.4}$$

$$a(S) = \frac{k\omega\,ko}{k\omega + ko} \frac{dPc}{dS} \tag{2.5}$$

Pc is the capillary pressure and $k\omega$, k_o the mobilities of the two fluids. The mobilities are related to relative permeabilities by:

$$k_i = \frac{k_{ri}}{\mu i} ; k_{ri} = \text{relative permeability}$$

To these three equations (2.1) to (2.3) an initial condition must be added:

$$S(x,o) = S \qquad\qquad (2.6)$$

and two boundary conditions.

Various laboratory experiments may be simulated by using appropriated boundary conditions. In all these equations (2.1) to (2.6), the unknowns are S and P. The other parameters $\Phi, K, \rho_\omega, \rho_o, \mu_o, \mu_\omega$ are constant and known, ko, kω and Pc are unknown curves that depend on the unknown S. ko, kω are related to water and oil relative permeabilities as indicated above.

The system of equations is approximated by a scheme similar to the one developed in previous references[5-6]. It uses a discontinuous piecewise linear slope limited approximation of the main unknown. A standard three-point scheme is used for the global pressure and the capillary diffusion term, second term in equation (2.1). In space, this scheme is of order two for the capillary diffusion term and of quasi order two for the fractional flow convection term. Its validity domain ranges from the pure diffusive case (natural imbibition) to the pure convective case (high velocity displacement), and yields in the latter case a reduced numerical diffusion as compared with the usual up-stream weighted finite differences. The time discretization was chosen to be explicit for the convective term, but implicit for the diffusion term, hence the stability condition for the scheme is only in $\Delta t / h$ (time step over space discretization).

3 - PARAMETER ESTIMATION METHOD

3.1. Formulation of a least squares problem

The following data are available from laboratory experiments :

- cumulative production volumes Q_k^m measured (superscript m) at different time k.

- pressure drop ΔP_k^m across the core measured at different times k

- variation $S_{k,i}^m$ of the saturation at ith measurement point against time k.

The measurements of saturation profiles $S_{k,i}^m$ offer extremely useful information for determining relative permeabilities and capillary pressure.

The numerical model and the experimental data are used to build up an error function associated with a choice of the parameter vector defining the relative permeability and capillary pressure curves

$$J(a) = W_Q \sum_k (Q_k^c - Q_k^m)^2 + W_P \sum_k (\Delta P_k^c - \Delta P_k^m)^2 \tag{3.1}$$

$$+ W_S \sum_k \sum_i (S_{k,i}^c - S_{k,i}^m)^2$$

where $Q_k^c, \Delta P_k^c$, and $S_{k,i}^c$ are computed (superscript c) using the numerical models. $Q_k^m, \Delta P_k^m$, and $S_{k,i}^m$ are the corresponding measured data and W_Q, W_P, and W_S are weighting factors to be chosen by the user. In this equation all the variables are dimensionless. Zero values are adopted for these weighting factors when the corresponding measurement are lacking.

3.2. Computing the gradient

The error function J(a) is minimized using an optimization routine[9-11]. As usual, this optimization routine requires two subroutines, one to compute J(a) and another to compute its gradient $\nabla J(a)$. A special feature of our code concerns the way these two routines are obtained. In previous works, the routine for the computation of $\nabla J(a)$ was obtained in one of two possible ways:

- $\nabla J(a)$ was computed by slightly varying the parameters one at a time and using a finite difference formula. This approach is easy to implement, but is computer time consuming. It only gives an approximation of the true gradients, and hence does not allow for a thorough minimization of the error function.

- $\nabla J(a)$ was computed exactly by using an adjoint equation[12-13] (optimal control theory). This approach is the most efficient in computing time (only one additional simulation per gradients evaluation) and in optimization efficiency (exact gradient), but is difficult to implement. In fact, the complexity of the equations in the forward model, the calculation (by hand) of the adjoint equation to such a model, its coding and its debugging until the calculated gradient agrees for all components with a carefully computed finite difference gradient, may take up to six months of manpower. Moreover, once this is achieved for one forward model, the whole process has to be repeated from scratch if one simply wants to change a boundary condition, or the discretization scheme, for example.

The approach proposed combines the implementation simplicity of the first approach with the computing performance of the second. It is based on the use of a Symbolic Code Generator (SCG)[14] developed at INRIA. This code, written in Macsyma, takes as input the discretized equations of the forward model and of the error function. Note that the full expertise concerning the choice of the discretization scheme is left to the user: the SCG is not an expert system for the discretization of partial differential equations. In fact, it does not know where the equations it is fed come from. Once the equations have been entered, and the user has specified which variables represent unknown parameters or given data, the SCG analyzes the equations, recognizes if they are explicit or implicit, in which case it checks for linearity (the user has then to propose few subroutines from a scientific library) or non linearity (the SCG proposes to solve the equations by Newton's method).Once these choices have been made, the SCG generates the Fortran code corresponding to the resolution of the discretized equations and the evaluation of the error function $J(a)$. It then determines the adjoint equation using symbolic calculation, and generates the corresponding Fortran code for its resolution and the evaluation of $\nabla J(a)$. Hence the output of ghe SCG is a Fortran subroutine for the computation of $J(a)$ and $\nabla J(a)$, which is the kernel of the code.

This approach enabled to develop our code in a very efficient and reliable way. The model described above was processed by running the SGC with the *ad hoc* discretized equations. The generation of the code required only a few days of manpower. The tests for the generated gradient code (comparison with a thoroughly computed finite different gradient) always succeeded the first time, confirming the reliability of the symbolic approach.

3.3. Representation of parameters

The reduction of the relative permeability and capillary pressure curves to finite dimensionality was done in two classic ways:
<u>analytical representation</u>: the usual empirical equations in the power function are used for relative permeabilities

$$kr_\omega(S)=a_\omega S^{b_\omega} \qquad (3.2)$$

$$kr_0(S)=a_0(1-S)^{b_0} \qquad (3.3)$$

where $a_\omega, b_\omega, a_0, b_0$ are the parameters to be estimated.
A number of closed form formulas depending on a few coefficients were used for capillary pressure curves.

<u>discrete representation</u>: the reduced [0,1] interval is discretized into NS intervals of length ΔS, and all saturation dependant curves are assumed to be piecewise linear continuous.

Note that changing the parametrization of unknown functions is a very simple operation. For example, using an other analytical representation, or a higher order spline discrete representation does not require rerunning the SCG generator. Once a code has been generated for the calculation of the gradient with respect to the discretized representation, a very simple matrix calculation yields the gradients with respect to the analytical representation.

The analytical solution form usually helps to find an approximate solution rapidly, while the discrete representation can be used to enhance the fitting with the recorded data.

4 - RESULTS

The simulator above described can be used to calculate production, pressure drop and saturation profiles at given times. These data can be considered as observations and the estimation method can be used to retrieve the relative permeabilities and the capillary pressure used to calculate them.

This was done and presented in[15]

In this section the estimation of relative permeabilities and capillary pressure is derived from the measurements performed during a laboratory water-oil imbibition displacement. The measurements consist of saturation profiles, pressure drop across the core and cumulative oil recovery. In this experiment, water is injected at a constant flow rate into a sample that initially contains oil and water. The core is placed in a vertical position to study the effect of gravity.

The following data are available: 10 experimental saturation profiles measured at 55 points of the core, pressure drop across the core at 30 different times, and cumulative oil recovery. The data in the saturation profiles (about 600 values) are more extensive than the other data.

Several calculation tests of relative permeabilities and capillary pressure were performed:

1. Relative permeabilities were estimated in the power function form, (5.1) and (5.2), with an experimental local capillary pressure. Results found were: $a_\omega = 0.038, a_0 = 1.2, b_\omega = 2.137, b_{n\omega} = 1.173$ after 20 iterations with initial values $a_\omega = 0.2, a_0 = 0.8, b_\omega = 2., b_{n\omega} = 2..$ The final error function was 100 times lower than the initial values.

The experimental data are plotted in Figures 1 and 2 together with the corresponding calculated values.

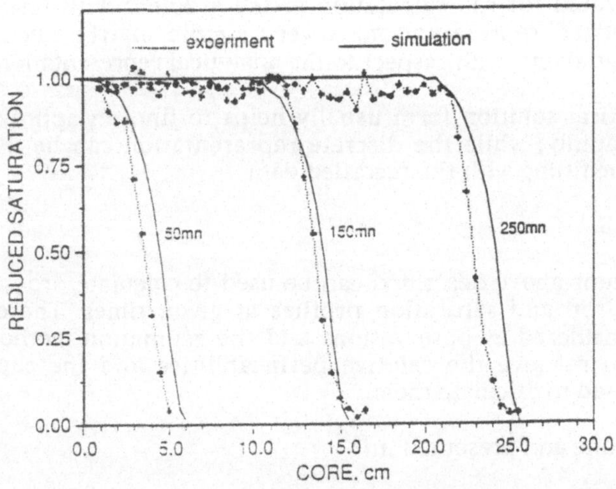

Figure 1 Comparison of measured and calculated saturation profiles with an experimental capillary pressure curve.

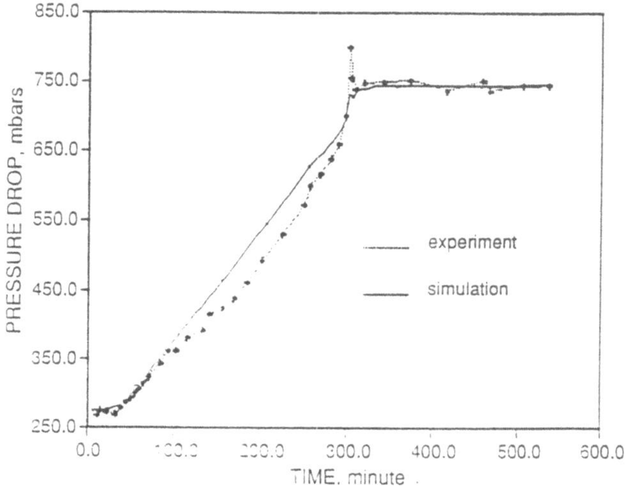

Figure 2 Comparison of measured and calculated pressure drop with an experimental capillary pressure curve.

2. Relative permeabilities in the power function form and the capillary pressure in piecewise linear form were estimated. Results found were: $a_\omega = 0.038, a_o = 1.2, b_\omega = 2.06, b_o = 1.16$. These values are very close to the previous ones. The capillary pressure is also quite well estimated as shown in Figure 3. The saturation profiles calculated with these kr, Pc are very similar to the experimental ones (Figure 4). The final error function is 150 times lower than the initial values.

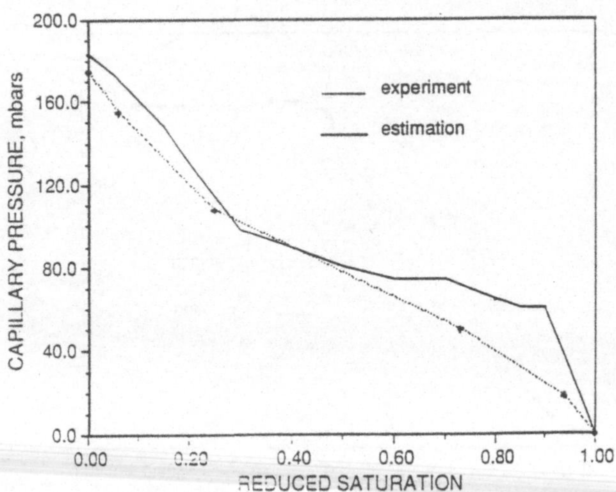

Figure 3 Comparison of experimental and estimated capillary pressure curves.

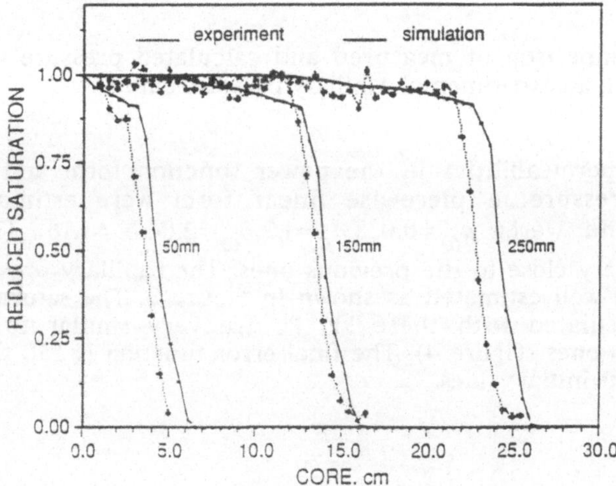

Figure 4 Comparison of measured and calculated saturation profiles with the estimated capillary pressure curve.

The results are satisfactory. A good match of saturation profiles and pressure drop of a laboratory experiment were obtained by using kr in power function form and Pc in piecewise linear function. The residual errors on saturation profiles and pressure drop curves can be partially explained by heterogeneities in the core sample, which are not taken into account by the numerical model.

5 - CONCLUSIONS

A method for simultaneously estimating relative permeabilities and capillary pressure from laboratory experiments has been described. By comparison with comparable methods[16], the solution of the equations, local saturations, is taken into account in the inverse problem.

The main characteristics of such an algorithm are as follows :

- It minimizes an error function measuring the error between measured observations and observations calculated by a numerical model.

- The gradient of the error function is calculated through techniques derived from optimal control theory.

- The two routines calculating the error function and its gradient are generated by a symbolic generator: SCG. This use of SCG decreases implementation time and improves reliability.

- A standard quasi-Newton method was used to solve the optimization problem.

The same approach was used for another kind of laboratory experiment: in centrifuge experiment[17].

6 - ACKNOWLEDGMENTS

This work and the proposed approach presented in this paper were the result of a very fruitful collaboration between IFP and INRIA (Le Chesnay, France) and specially Professor Guy Chavent, Jerôme Jaffre and Jun Liu.

7 - NOMENCLATURE

S	normalized saturation
P	global presure
ΔP	pressure drop across core (dimensionless)

Q	cumulative production (dimensionless)
kr_ω	water relative permeability
kr_o	oil relative permeability
Pc	capillary pressure (Pa)
a_ω, a_o	end-point values of relative permeability curves
b_ω, b_o	exponents in power relative permeability models
ρ_ω, ρ_o	density
M_ω, M_o	viscosity
K	rock permeability
Φ	rock porosity
k_ω, k_o	mobility
g	gravity
$J(a)$	error function
$\nabla J(a)$	gradient of J(a)
W_P	weighting factor for pressure drop (dimensionless)
W_Q	weighting factor for production (dimensionless)
W_S	weighting factor for saturations (dimensionless)
NS	number of discretized points for estimated curves

Superscrit

m	measured
c	calculated

Subscript

ω	water
o	oil
k	index of measured time points
i	index of measured space points
j	discretization subscript for estimated curves

8 - REFERENCES

[1] Chavent, G., Cohen, G., and Espy, M.: "Determination of Relative Permeabilities and Capillary Pressures by an Automatic Adjustment Method", paper SPE 9237 presented at the 1980 SPE Annual Technical Conference and Exhibition, Dallas, Sept. 21-24.

[2] Kerig, P.D. and Watson, A.T.: "A New Algorithm for Estimating Relative Permeabilities From Displacement Experiments", SPEJ (Feb. 1987) 103-12, Trans., AIME, 283.

[3] Yang, P.H. and Watson, H.T.: "A Bayesian Methodology for Estimating Relative Permeability Curves", SPERE (May 1991) 259-65.

[4] Richmond, P.C. and Watson, H.T.: "Estimating Multiphase Flow Functions From Displacement Experiments", paper presented at the 1988 Spring Natl. Meeting, Chemical Engineering Problems in Flow in Porous Media, March 6-10.

[5] Chavent, G. and Jaffre, J.: Mathematical Models and Finite Elements for Reservoir Simulation", North Holland (1986) 376.

[6] Van Leer, B.: "Towars the Ultimate Conservative Scheme: IV. A New Approach to Numerical Convection", J. Comp. Phys. (1977) 23, 276-99.

[7] Chavent, G.: "Identifiability of Parameters in the Output Least Squares Formulation", Structural Identifiability of Parametric Models, E. Walter (ed.), Pergamon Press, Elmsford, NY (1986).

[8] Chavent, G., Dupuy, M., and Lemonnier, P.:"History Matching by Use of Optimal Control Theory", SPEJ (Feb. 1975) 74-86; Trans. AIME, 259.

[9] Chavent, G.: Identification of Functional Parameters in Partial Differential Equations, in Identification of Parameters in Distributed Systems, Goodson R.E. and Polis M., New-York (1974) 31-48.

[10] Chavent, G.: Identifiability of Parameters in the Output Least Square Formulation, in Structural Identifiability of Parametric Models, Walter E., Pergamon (1986).

[11] Chavent, G.: New Trends in Identification of Distributed Parameters Systems, to appear in the Proc. of the 4th IFAC Congress, Munich, July 27-31 1987.

[12] Chavent, G., Cohen, G., and Espy, M.: Determination of relative Permeabilities and Capillary Pressures by an Automatic Adjustment Method, paper SPE 9237 presented at the 1980 Annual Technical Conference and Exhibition, Dallas, Sept. 21-24.

[13] Watson, A.T., Seinfeld, J.H. and Gavalas, G.R. History Matching in Two-phase Petroleum Reservoirs, paper SPE 8250 presented at the 1979 Annual Fall technical Conference and Exhibition, Las Vegas, Sept. 23-26.

[14] Liu, J.: Calcul Formel et Probleme inverse PhD Thesis, University of Paris-Dauphine (1989).

[15] Chardaire-Rivière, C., Chavent, G., Jaffré, J., Liu, J., Bourbiaux, B. Simultaneous Estimation of Relative Permeabilities and Capillary Pressure. SPE Formation Evaluation, December 1992, 283-289.

[16] Richmond, P.C. and Watson, H.T. Estimating Multiphase Flow Functions from Displacement Experiments. Paper presented at the 1988 Spring Natl. Meeting, Chemical Engineering Problems in Flow in Porous Media, March 6-10.

[17] Chardaire-Rivière, C., Forbes, P., Feng Zhang, J., Chavent, G., Lenormand, R. Improving the Centrifuge Technique by measuring local saturations. Paper SPE 24882, presented at the 67th Annual Technical Conference and Exhibition of the SPE, Washington, DC, October 4-7.

Catherine Rivière
Institut Francais du Pétrole
1 et 4, avenue de Bois - Préau
92 Rueil - Malmaison   FRANCE

Mathematical Methods for 2D Reflector Design

Maurice Maes

Philips Research Laboratories
P.O.Box 80.000 – 5600 JA Eindhoven
The Netherlands

1   Introduction

A basic problem in lighting technology is to design an optical system that illuminates a certain object in a prescribed manner. If this optical system consists of a fixed light source and a reflector that is to be chosen, then this problem is called a *reflector design problem*. This paper discusses mathematical methods which help to solve the reflector design problem.

In its general form, the reflector design problem is too comprehensive. Therefore, one first has to concentrate on special classes of problems that allow mathematical treatment. We mention several factors that in practice are to be dealt with, and which can have a strong impact on the difficulty of the problem.

(a) *The light source.* For mathematical treatment, it is a great advantage if one can assume the light source to be a point source, since then the luminous intensity of the source is easy to describe.

(b) *Symmetry.* The general reflector problem is of course a 3-dimensional problem. In some cases however, viz. in a cylindrically symmetric situation with a linear light source or in a rotationally symmetric situation with a point source, a 2-dimensional approach is possible.

(c) *The material of the reflector.* Reflection may be specular or diffuse, and there may be more or less absorption, depending on the material of the reflector.

(d) *Near and far fields.* Generally, the prescribed illumination will be given on a flat screen at a fixed distance from the optical system. In this case, the problem will be called a *near field problem.* A typical example is the design of an LCD-backlighter. In some illumination tasks, the distance between the optical system and the object is that large that the required illumination may as well be described as a luminous intensity distribution on the *angles* of the reflected rays, and we will speak of a *far field problem.*

Naturally, these factors are not the only ones that make the general problem so difficult: in practice, multiple reflections may occur, reflected rays may re-enter the source, design specifications may put restrictions on the dimensions of the reflector, etc.

As far as the items above are concerned, the scope of this paper is the following. We will only consider the cases of linear light sources in cylindrically symmetric situations and point sources in rotationally symmetric situations. These cases can be reduced to a 2-dimensional problem. Reflection is assumed to be specular. The first results towards the general solution to the 2-dimensional problem that we consider in this report have been found in 1958 by Keller [6], and also by Trembač [13].

The strategy to solve the problem, which generally has infinitely many solutions, is as follows. First one establishes a relation between incident and reflected rays that produces the required distribution. Then one calculates the reflector that realizes this relation by complying to the law of reflection. From the mathematical viewpoint, this second step is very straightforward. The interesting part of the design method is the first step. The freedom of choice of the relation between incident and reflected rays can be used for instance to influence the dimensions of the reflector or to avoid interaction of reflected rays with the light source. Unfortunately, this aspect has received little attention in the literature so far.

It is our aim to illustrate that successful application of mathematics to reflector design involves much more than only solving differential equations. We illustrate this by considering the 2-dimensional problem, which is introduced in Section 2. The canonical, monotonic solutions to the problem are (for far fields) given in Section 3. These solutions will in general not directly be of help to a designer, but they form the building blocks of the subdivided solutions introduced in Section 4, which may be of real practical help. Some brief remarks on 3D problems will end that section.

2 The Two-Dimensional Problem

2.1 Terminology

There is no need for a detailed description of the few notions from lighting
theory that will be used in this paper. Most of these are intuitively clear.
Moreover, they will be used in a 2-dimensional and mathematical context,
so it is of no use to specify units. Let us just mention the essentials, which
are defined in accordance with Keitz [5] to which the reader is referred for
more details.

In the case of a far field problem, the required distribution will be called
the *(luminous) intensity distribution $\tilde{E}$*, which is the limiting value of the
luminous flux per solid angle, with the solid angle approaching zero. The
notion of luminous intensity can also be used to describe the radiation I of
point sources. In the case of a near field problem, the required distribution
on the screen will be called the *illumination E*, which is the luminous flux per
area of the surface thus illuminated. Now, the *inverse square law* describes
the relation between luminous intensity and illumination in for point sources:
"*the illumination at a point in a plane perpendicular to the line joining the
point and the source is equal to the luminous intensity of the source in the
direction of the point, divided by the square of the distance between point and
source*" [5][Ch. VI, p. 80], i.e.

$$E = \frac{I}{d^2},\tag{1}$$

where I is the luminous intensity of the source and d is the distance between
the point and the source. In the more general case in which the light falls in
at an angle α to the normal to the plane, we find

$$E = \frac{I \cos \alpha}{d^2}.\tag{2}$$

Throughout the report, 3-dimensional intensities and illuminations will be
denoted I, E and $\tilde{E}$, while 2-dimensional intensities and illuminations will
be denoted $\mathcal{I}$, $\mathcal{E}$ and $\tilde{\mathcal{E}}$. In far field problems, the required distribution is
defined on angles; in order to distinguish these problems notationally from
near field problems, we use a 'tilde' and write $\tilde{E}$ and $\tilde{\mathcal{E}}$.

2.2   The Rotationally Symmetric Case

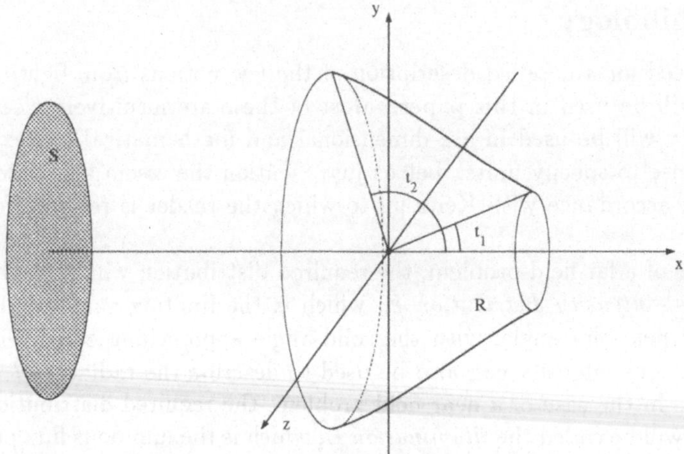

Figure 1: The rotationally symmetric case.

In this section we describe the rotationally symmetric 3-dimensional model problem which allows a 2-dimensional treatment. Consider Fig. 1. Suppose we have a rectangular xyz-coordinate system in $\mathbf{R}^3$, with a point source located at the origin with radiation which is rotationally symmetric around the x-axis. Let the luminous intensity of the source in directions

$$(\cos t, \sin t \cos \phi, \sin t \sin \phi)$$

of the reflector be defined by a function $I(t, \phi)$. Rotational symmetry of the distribution of the source means that $I(t, \phi)$ is independent of ϕ, so we will write $I(t) := I(t, \phi)$.

Furthermore, suppose we have fixed angles $t_1, t_2 \in [0, \pi)$ with $t_1 < t_2$. Then a rotationally reflector around the x-axis between angles t_1 and t_2 can be described by a surface of the form

$$R := \{(r(t) \cos t, r(t) \sin t \cos \phi, r(t) \sin t \sin \phi) \in \mathbf{R}^3 \mid t \in [t_1, t_2], \ \phi \in [0, 2\pi]\}, \tag{3}$$

where r is a continuous positive function.

Also, suppose that we have a screen that is represented by a disk S of the form

$$S := \{(x, y, z) \in \mathbf{R}^3 \mid x = -h, \ y^2 + z^2 \leq y_2^2\} \tag{4}$$

for a fixed distance h, and for fixed radius y_2.

Now let a required illumination E be defined on S which is constant on circles in S with center $(-h, 0, 0)$, so E is of the form

$$E(y, z) = e(y^2 + z^2) \tag{5}$$

for some function $e : [0, y_2] \to \mathbf{R}^+$. We will assume that e is a non-negative, integrable function. We may assume that E is the required illumination *for the reflected light only*. We will consider the problem of determining a reflector of the form (3) that realizes this required illumination E.

2.3 The Cylindrically Symmetric Case

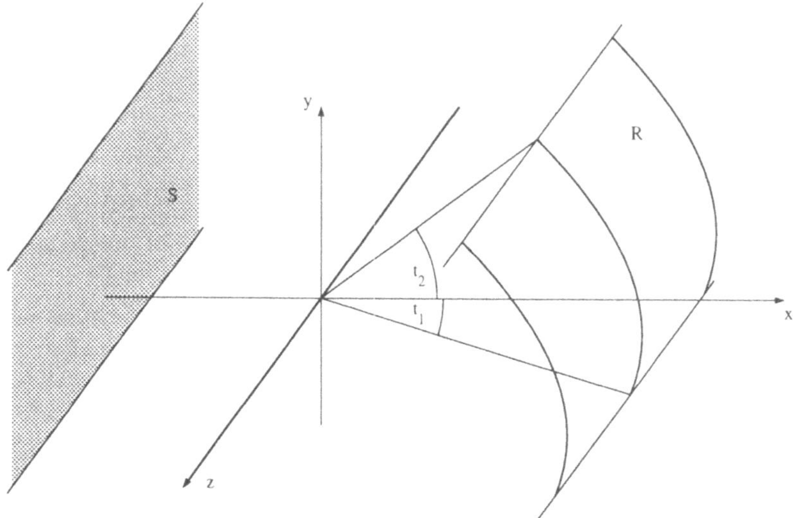

Figure 2: The cylindrically symmetric case.

In this section we describe the cylindrically symmetric 3-dimensional model problem which allows a 2-dimensional treatment. Consider Fig. 2. Suppose that we have a rectangular xyz-coordinate system in $\mathbf{R}^3$, and that we have a linear, infinitely long light source l that coincides with the z-axis, of uniform radiation. This means that a line element dl at position $(0,0,s)$ can be seen as a point source with intensity $I(t,\chi)$ in directions $(\cos t \sin \chi, \sin t \sin \chi, \cos \chi)$ of the reflector. The uniformity of l implies that this intensity distribution is independent of s.

Furthermore, suppose we have fixed angles $t_1, t_2 \in (-\pi, \pi)$ with $t_1 < t_2$. Then a cylindrical reflector for l between angles t_1 and t_2 can be described by a surface of the form

$$R := \{(r(t)\cos t, r(t)\sin t, z) \in \mathbf{R}^3 \mid t \in [t_1, t_2]\}, \tag{6}$$

where r is a continuous positive function.

Now, suppose that we have an infinite strip S (representing the screen) given by

$$S := \{(x,y,z) \in \mathbf{R}^3 \mid x = -h, \ y_1 \leq y \leq y_2\} \tag{7}$$

for fixed distance h and heights y_1 and y_2.

Finally, we assume we have a required illumination E on S which is independent of z, so E is of the form

$$E(y,z) = e(y) \tag{8}$$

for some function $e : [y_1, y_2] \to \mathbf{R}^+$. We again assume that e is a non-negative, integrable function. We may assume that E is the required illumination *for the reflected light only*. We will consider the problem of determining a reflector of the form (6) that realizes this required illumination E.

2.4   Geometry of the Two-Dimensional Problem

We will now introduce some notation for the 2-dimensional problem; see Fig. 3. We assume a point source to be located in the origin of a rectangular xy-coordinate system. Where angles of rays are concerned, there is a subtlety in the notation that we will use. Those rays that are incident on the reflector will be denoted by Arab letters, usually t or s, and they are measured clockwise relative to the positive direction of the x-axis. Reflected rays, and those emitted rays that are not incident on the reflector, will be denoted by Greek letters, usually θ or ψ, and they are measured counterclockwise relative to the negative direction of the x-axis.

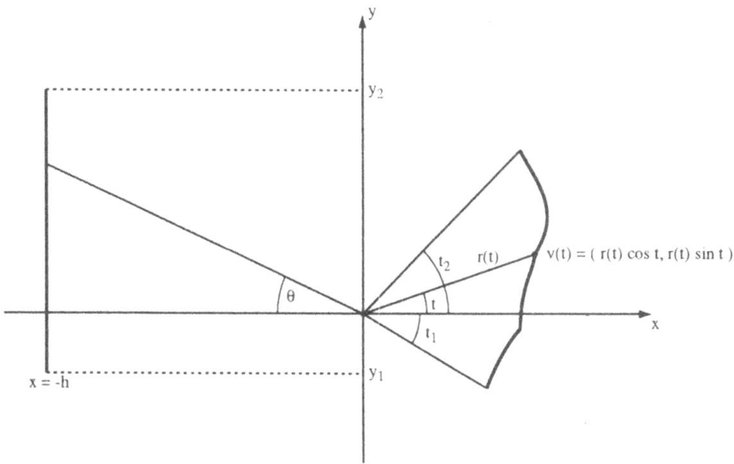

Figure 3: The geometry of the 2-dimensional problem.

We can describe the reflector by a function $r : [t_1, t_2] \to \mathbf{R}$, where the reflector is located between angles t_1 and t_2, with $t_1 < t_2$, and where $r(t)$ is the distance from the source to the reflector in direction t. The corresponding reflector point will be denoted $v(t)$, i.e.

$$v(t) := (r(t)\cos t, r(t)\sin t), \tag{9}$$

so the reflector corresponds to the curve $v : [t_1, t_2] \to \mathbf{R}^2$. In the case of a near field problem, the screen will be represented by a segment of the line $x = -h$ between heights y_1 and y_2 ($y_1 < y_2$).

The direction of a ray that is reflected from the point $r(t)$, will be denoted $\theta(t)$. Consider Fig. 4 in which the (two-dimensional) law of reflection is illustrated. This law, that states that incident and reflected rays form equal angles with the normal to the reflector at the point of incidence, can be formulated as

$$\frac{\dot{r}(t)}{r(t)} = \tan\left(\frac{t + \theta(t)}{2}\right), \tag{10}$$

where we assume that the derivative $\dot{r}(t)$ exists.

The height at which the reflected ray enters the screen will be denoted

130

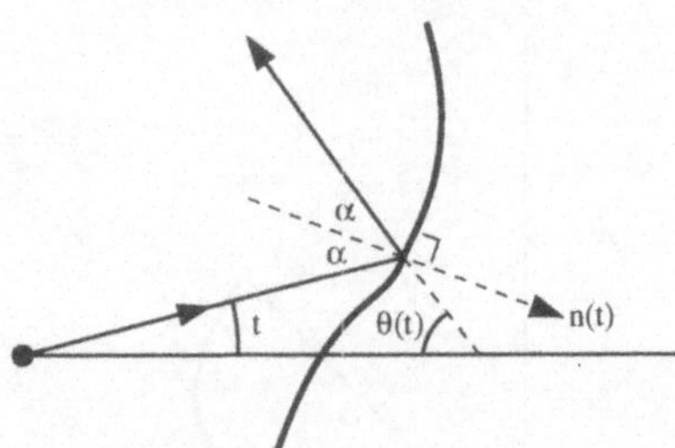

Figure 4: Illustration of the law of reflection.

$y(t)$. The relation between $y(t)$ and $\theta(t)$ in terms of t and r is as follows:

$$\frac{y(t) - r(t)\sin t}{h + r(t)\cos t} = \tan \theta(t). \qquad (11)$$

Combining (10) and (11) gives $y(t)$ expressed in terms of t and r. We get

$$\frac{\dot{r}(t)}{r(t)} = \tan\left(\frac{t}{2} + \frac{1}{2}\arctan(\frac{y(t) - r(t)\sin t}{h + r(t)\cos t})\right). \qquad (12)$$

2.5 Problem Statement

In this section we describe the general 2-dimensional problem that we are interested in. Both the cylindrically and rotationally symmetric problems as described above can be reduced to this 2-dimensional problem.

Let $\mathcal{E} : [y_1, y_2] \to \mathbf{R}^+$ be a non-negative integrable function which describes the required illumination for the reflected light on the screen that is represented by the interval $[y_1, y_2]$ on the vertical $x = -h$. Also, let $\mathcal{I} : [t_1, t_2] \to \mathbf{R}^+$ be a non-negative integrable function which describes the intensity of the light source in the directions of the reflector. We may assume that $\mathcal{I}$ accounts for the *reflection coefficient* ρ of the reflector material as well, assuming that this coefficient is constant.

Furthermore, we assume that we have

$$\int_{y_1}^{y_2} \mathcal{E}(u)\,du = \int_{t_1}^{t_2} \mathcal{I}(s)\,ds. \qquad (13)$$

This condition may be thought of as the condition of conservation of energy, and it is necessary for the problem that we will consider to have a solution.

Now, informally speaking, the problem that we consider in this paper is the following.

Problem 2.1. *Consider the geometric situation described in Section 2.4, and let $\mathcal{E}$ and $\mathcal{I}$ be as defined above, satisfying (13). Find a reflector $r(t)$ between angles t_1 and t_2, such that, in combination with a point source of distribution $\mathcal{I}$, it produces the illumination $\mathcal{E}$ on the interval $[y_1, y_2]$.*

We will now illustrate how the two 3-dimensional problems of Sections 2.2 and 2.3 can be reduced to this 2-dimensional problem. We will start with the rotationally symmetric case.

We have a required distribution for the reflected light on the disk S given by (4), of the form $E(y, z) = e(y^2 + z^2)$. The intensity distribution of the source in the directions of the reflector is given by the function $I(t, \phi)$, and we may assume, as in the previous section, that this function has already been corrected for the reflection coefficient of the reflector material. The design problem has a solution only if

$$\int_S E(y, z)\, dy\, dz = \int_\Omega I(t, \phi)\, d\omega. \tag{14}$$

Here Ω is the set of directions of those rays that meet the reflector R (given by (3)), so $\Omega = \{(t, \phi)|t_1 \le t \le t\}$. We can write $d\omega = \sin t\, dt\, d\phi$. After substituting $y = u \cos \phi$, $z = u \sin \phi$, such that $dy\, dz = u\, du\, d\phi$, we find

$$\int_0^{2\pi} \int_0^{y_2} e(u^2)u\, du\, d\phi = \int_0^{2\pi} \int_{t_1}^{t_2} I(t, \phi) \sin t\, dt\, d\phi, \tag{15}$$

so

$$\int_0^{y_2} e(u^2)u\, du = \int_{t_1}^{t_2} I(t) \sin t\, dt. \tag{16}$$

The following result is a consequence of this relation.

Proposition 2.2. *Consider the 2-dimensional problem of the previous section with*

$$\mathcal{E}(u) = e(u^2)u \qquad \textit{for all } u \in [0, y_2], \textit{ and}$$
$$\mathcal{I}(t) = I(t) \sin t \qquad \textit{for all } t \in [t_1, t_2].$$

Now if $r(t)$ is a curve that solves the 2-dimensional problem for this $\mathcal{E}$ and $\mathcal{I}$, then the reflector surface

$$R := \{(r(t) \cos t, r(t) \sin t \cos \phi, r(t) \sin t \sin \phi) \in \mathbf{R}^3 \mid t \in [t_1, t_2], \ \phi \in [0, 2\pi]\}$$

solves the 3-dimensional problem for E and I as above.

For the cylindrically symmetric case of Section 2.3, conservation of energy per unit length is easily seen to be equivalent to

$$\int_{y_1}^{y_2} e(u)\, du = \int_{t_1}^{t_2} I^*(t)\, dt, \tag{17}$$

where $I^*(t) := \int_0^\pi I(t, \chi) \sin \chi\, d\chi$, and we obtain the following result.

Proposition 2.3. *Consider the 2-dimensional problem of the previous section with*

$$\begin{aligned}
\mathcal{E}(u) &= e(u) && \text{for all } u \in [y_1, y_2], \text{ and} \\
\mathcal{I}(t) &= \textstyle\int_0^\pi I(t, \chi) \sin \chi\, d\chi && \text{for all } t \in [t_1, t_2].
\end{aligned}$$

Then if $r(t)$ is a curve that solves the 2-dimensional problem for this $\mathcal{E}$ and $\mathcal{I}$, then the reflector surface

$$R := \{(r(t) \cos t, r(t) \sin t, z) \in \mathbf{R}^3 \mid t \in [t_1, t_2]\}$$

solves the 3-dimensional problem for E and I as in the cylindrically symmetric case of Section 2.3.

The proofs of both propositions are omitted; they are fairly straightforward.

2.6 From Near Fields to Far Fields

It may happen that the distance h from the light source to the screen is much larger than the dimensions of the reflector to be designed. (Street or playground lighting are typical applications in which we have such a situation.) In that case we usually may describe the required illumination as an intensity distribution on the *angles* of the reflected rays. This intensity distribution can then be defined in such a way, that, if all light would come directly from the light source, the correct illumination would be accomplished. Since the dimensions of the reflector are small compared to h, we may expect that the required illumination is approximated sufficiently accurately when the rays are reflected from the reflector rather than directly from the source. The reason why we would want to consider the far field problem is that the computation of the reflector and the prediction of some of its properties is easier for far than for near field problems.

For both rotationally and cylindrically symmetric problems, we can formulate the 3-dimensional far field problems, and then deduce a corresponding 2-dimensional problem. We may also start with the general 2-dimensional near field problem, and then deduce the corresponding far field problem. The latter gives the same result, and is somewhat easier.

Consider the required illumination $\mathcal{E}(u)$ on the interval $[y_1, y_2]$ on the vertical at distance h. On a small line element du on this interval, an amount of light $\mathcal{E}(u)\,du$ is required. Now, we may write $u = h\tan\psi$, and $du = h/\cos^2\psi\,d\psi$, so

$$\mathcal{E}(u)\,du = \mathcal{E}(h\tan\psi)\frac{h}{\cos^2\psi}\,d\psi. \tag{18}$$

Let $\theta_1 := \arctan(y_1/h)$ and $\theta_2 := \arctan(y_2/h)$. Then if the illumination $\mathcal{E}$ was to be realized by a light source located at the origin, it would have to produce the intensity distribution

$$\tilde{\mathcal{E}}(\psi) := \mathcal{E}(h\tan\psi)\frac{h}{\cos^2\psi}. \tag{19}$$

So, what we can do is prescribe the intensity distribution (19) for the reflected light. Note that we still have conservation of energy:

$$\int_{\theta_1}^{\theta_2} \tilde{\mathcal{E}}(\psi)\,d\psi = \int_{t_1}^{t_2} \mathcal{I}(s)\,ds. \tag{20}$$

In the rotationally symmetric case, we have

$$\tilde{\mathcal{E}}(\psi) = e(h^2\tan^2\psi)\frac{h^2\sin\psi}{\cos^3\psi} \tag{21}$$

and in the cylindrically symmetric case, we have

$$\tilde{\mathcal{E}}(\psi) = e(h\tan\psi)\frac{h}{\cos^2\psi}. \tag{22}$$

3 Monotonic Solutions for Far Field Problems

3.1 The Main Strategy

In this chapter we will consider the main strategy to solve the reflector design problem for far fields only. We have a required intensity distribution $\tilde{\mathcal{E}}$ that is defined on *angles*, let's say on an interval $[\theta_1, \theta_2]$ with $\theta_1 < \theta_2$. Also, we have a function $\mathcal{I} : [t_1, t_2] \to \mathbf{R}$ describing the luminous intensity of the source in the directions of the reflector.

The problem of designing a reflector between angles t_1 and t_2 for the above problem can of course only be solved if there is conservation of energy, i.e. if

$$\int_{\theta_1}^{\theta_2} \tilde{\mathcal{E}}(\psi)\, d\psi = \int_{t_1}^{t_2} \mathcal{I}(s)\, ds. \tag{23}$$

In order to present the global strategy to solve the problem, let us first assume that we have a reflector that realizes the required distribution. For each emitted ray t, let $\theta(t)$ be the direction of the reflected ray. We may describe this relation by a function $\theta : [t_1, t_2] \to [\theta_1, \theta_2]$. From (10) it then follows that the reflector is described by the function

$$r_\theta(t) = r_\theta(t_1) \exp\left(\int_{t_1}^{t} \tan\left(\frac{s + \theta(s)}{2}\right) ds\right). \tag{24}$$

Here we write $r_\theta(t)$ to emphasize that r depends on the choice of θ (and we will see later on that many feasible θ's may exist).

Now, consider an angular interval $[\psi_1, \psi_2] \subset [\theta_1, \theta_2]$, with $\psi_1 < \psi_2$. The required amount of light within $[\psi_1, \psi_2]$ equals

$$\int_{\psi_1}^{\psi_2} \tilde{\mathcal{E}}(\psi)\, d\psi.$$

Let $S_\theta[\psi_1, \psi_2]$ be the set of directions of emitted rays that are reflected into the interval $[\psi_1, \psi_2]$, i.e.

$$S_\theta[\psi_1, \psi_2] = \{t \mid \psi_1 \le \theta(t) \le \psi_2\}. \tag{25}$$

The corresponding amount of reflected light is equal to

$$\int_{S_\theta[\psi_1, \psi_2]} \mathcal{I}(s)\, ds.$$

So we find that

$$\int_{\psi_1}^{\psi_2} \tilde{\mathcal{E}}(\psi)\, d\psi = \int_{S_\theta[\psi_1,\psi_2]} \mathcal{I}(s)\, ds, \qquad (26)$$

for all ψ_1, ψ_2.

We see that if there is a reflector $r(t)$ that solves the problem, then the corresponding relation θ between incident and reflected rays satisfies (24) and (26). But the converse is also true! Each function θ that satisfies (26) for all ψ_1 and ψ_2 defines a relation between incident and reflected rays that will produce the required luminous intensity distribution, if there is a reflector that accomplishes this relation. Now, by (24) there is a function $r(t)$ that suits our purposes in the sense that the corresponding reflector reflects the rays into the required directions, provided that multiple reflections do not occur.

In the following sections we will see how to find functions θ that satisfy (26). The choice of θ is an important design step, because it will influence the shape of the reflector. Moreover, it may be used to avoid reflected rays re-entering the light source (if the finite dimensions of the source are taken into account) or to influence the angles of incidence on the screen, which may be important for near field problems.

From (10) it follows that if θ is continuous, then $\dot{r}$ is continuous, i.e. the reflector is smooth. In the following sections we will see that usually there are two (and only two) solutions such that θ is even differentiable: the unique increasing and decreasing θ's which correspond to divergent and convergent ray bundles, respectively.

3.2 Monotonic Solutions

Monotonic solutions (i.e. solutions with θ either increasing or decreaseing) are the ones that are usually discussed in the literature. To see how they can be found, let us first assume that we have such a solution, and then deduce the conditions it satisfies.

We start with the increasing solution, i.e. with an increasing function θ^+. Note that if θ^+ is increasing, then $\theta^+(t_1) = \theta_1$ and $\theta^+(t_2) = \theta_2$, and no two reflected rays intersect, so the ray bundle is divergent. Since θ^+ realizes the required distribution, we must have

$$\int_{\theta_1}^{\theta^+(t)} \tilde{\mathcal{E}}(\psi)\, d\psi = \int_{t_1}^{t} \mathcal{I}(s)\, ds. \qquad (27)$$

Because $\tilde{\mathcal{E}}$ is integrable and positive, it has a strictly increasing primitive $\tilde{F}$ on $[\theta_1, \theta_2]$ with $\tilde{F}(\theta_1) = 0$. Also, $\mathcal{I}$ has a primitive G on $[t_1, t_2]$ with $G(t_1) = 0$. So we can write

$$\tilde{F}(\theta^+(t)) = G(t). \tag{28}$$

Now, since $\tilde{F}$ is strictly increasing, it has an inverse $\tilde{F}^{-1}$, and we find

$$\theta^+(t) = \tilde{F}^{-1}(G(t)). \tag{29}$$

This uniquely defines θ^+ in terms of $\tilde{\mathcal{E}}$ and $\mathcal{I}$.

The decreasing case, corresponding to a convergent ray bundle, is treated similarly. Let's say we have a decreasing solution θ^-. We then find

$$\int_{\theta^-(t)}^{\theta_2} \tilde{\mathcal{E}}(\psi)\, d\psi = \int_{t_1}^{t} \mathcal{I}(s)\, ds, \tag{30}$$

so

$$\tilde{F}(\theta_2) - \tilde{F}(\theta^-(t)) = G(t). \tag{31}$$

By (23) and by definition of $\tilde{F}$ and G, we have $\tilde{F}(\theta_2) = G(t_2)$, so

$$\theta^-(t) = \tilde{F}^{-1}(G(t_2) - G(t)), \tag{32}$$

and this uniquely defines θ^- in terms of $\tilde{\mathcal{E}}$ and $\mathcal{I}$.

Example 3.1. Let $\mathcal{I}(t) = 1$ for all $t \in [t_1, t_2]$. This corresponds to a cylindrically symmetric case with a uniform light source. In that case, we have $G(t) = t - t_1$ for all $t \in [t_1, t_2]$, so $\theta^+(t) = \tilde{F}^{-1}(t - t_1)$, and $\theta^-(t) = \tilde{F}^{-1}(t_2 - t)$. Let us require a uniform distribution of the reflected light between angles $\tilde{\mathcal{E}}(\psi) = 4$ for all $\psi \in [\theta_1, \theta_2] = [-\pi/6, \pi/6]$. Suppose we have $[t_1, t_2] = [-2\pi/3, 2\pi/3]$ and the boundary condition $r(-2\pi/3) = 1$. We then find

$$
\begin{aligned}
\tilde{F}(\psi) &= 4\psi + 2\pi/3 \\
\tilde{F}^{-1}(s) &= s/4 - \pi/6 \\
\theta^+(t) &= \tilde{F}^{-1}(t + 2\pi/3) = t/4 \\
\theta^-(t) &= -t/4
\end{aligned}
$$

Explicit expressions for $r_{\theta^+}(t)$ and $r_{\theta^-}(t)$ can now be derived from (24). The corresponding reflectors are shown in Fig. 5.

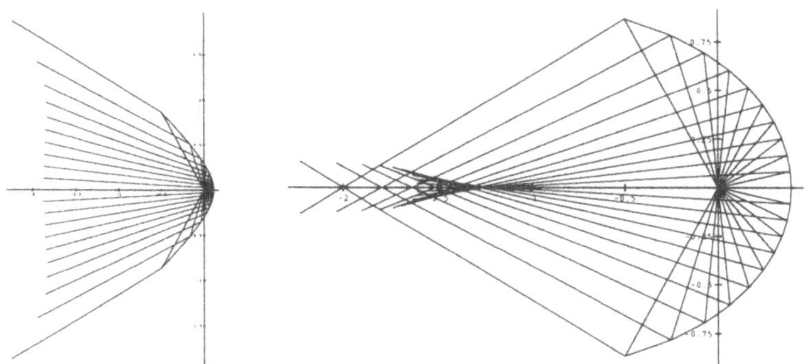

Figure 5: The monotonic solutions to Example 3.1.

3.3 The Direct Problem and Infinite Intensities

Now that we know the monotonic solutions to the reflector design problem, it is time to consider the so-called 'direct problem', i.e. the problem to determine the intensity distribution $\tilde{\mathcal{E}}$ when the reflector r is known. To that end, we have to divide the reflector into parts that yield monotonic ray paths, and simply add the contributions of all these parts.

Let us suppose we have a reflector described by a function r which is *twice* differentiable in all but at most a finite number of points. Then we find from (10) that

$$\theta(t) = 2\arctan\left(\frac{\dot{r}(t)}{r(t)}\right) - t, \tag{33}$$

and

$$\dot{\theta}(t) = \frac{2\ddot{r}r - 3\dot{r}^2 - r^2}{r^2 + \dot{r}^2}, \tag{34}$$

where we write r instead of $r(t)$, etc.

Suppose that θ is increasing on an interval $U = [u_1, u_2]$. Let the contribution of that interval to the intensity distribution be called $\tilde{\mathcal{E}}_U$. We have

$$\int_{\theta(u_1)}^{\theta(t)} \tilde{\mathcal{E}}_U(\psi)\, d\psi = \int_{u_1}^{t} \mathcal{I}(s)\, ds \tag{35}$$

for all $t \in U$. Differentiating (35) with respect to t gives

$$\tilde{\mathcal{E}}_U(\theta(t))\, \dot{\theta}(t) = \mathcal{I}(t), \tag{36}$$

138

so

$$\tilde{\mathcal{E}}_U(\theta(t)) = \frac{\mathcal{I}(t)}{\dot{\theta}(t)}. \tag{37}$$

Similarly, we find $\tilde{\mathcal{E}}_U(\theta(t)) = -\mathcal{I}(t)/\dot{\theta}(t)$ if θ is decreasing on U. Note that if $\mathcal{I}(t) = 1$ for all t, i.e. if the source has a uniform distribution, then

$$\tilde{\mathcal{E}}_U(\theta(t)) = |\dot{\theta}(t)|^{-1}. \tag{38}$$

It follows that if $\dot{\theta}(t) \to 0$, then $\tilde{\mathcal{E}}_U(\theta(t)) \to \infty$. Consequently, we find that if θ is a differentiable solution of a reflector design problem as considered above, where infinite intensities are not allowed, then θ must either be strictly increasing or strictly decreasing.

3.4 Applicability of Monotonic Solutions

The results of the previous sections seem to indicate that 2-dimensional reflector design problems with small sources are almost completely solved; especially from the mathematical viewpoint.

From the practical viewpoint however, a lot of work still has to be done. Practical designers do not have problems that are formulated in a mathematically correct sense, as defined above. Furthermore, in practice facetted reflectors are often to be preferred above smooth reflectors, for reasons of stability or manufacturing.

Also, the reflectors obtained by the methods above may not meet all kinds of other practical restraints, such as geometrical constraints, reflected rays avoiding the light source, etc.

Part of these problems can be overcome if we allow the raypaths to be discontinuous, that is if we allow the reflector to be non-smooth. In the following section we will see how infinitely many different solutions can be found if we allow non-smooth reflectors.

4 General Solutions By Subdivisions

In the previous section, we have introduced and discussed the monotonic solutions to the reflector design problem. The applicability of the method we have discussed so far would have been very limited if the two monotonic solutions were the only ones that could be found this way. In this section we will see that the monotonic solutions can be used as the 'building blocks' to (infinitely) many other solutions which might be more suitable to meet all kinds of practical requirements.

4.1 Subdividing the Reflector

Suppose we have a cylindrical problem with $\mathcal{I} = 1$ on $[t_1, t_2]$ (again for convenience only!) and that we have a required distribution $\tilde{\mathcal{E}}$ defined on $[\theta_1, \theta_2]$ as in the previous sections. Then we can subdivide the interval $[t_1, t_2]$ into n parts, as follows. Let

$$t_1 = u_0 < u_1 < \ldots < u_{n-1} < u_n = t_2$$

and let

$$U_i := [u_{i-1}, u_i]. \tag{39}$$

Then $\cup U_i = [t_1, t_2]$. At the same time, we might 'subdivide' the required distribution into n parts. Let $\tilde{\mathcal{E}}_1, \ldots, \tilde{\mathcal{E}}_n : [\theta_1, \theta_2] \to \mathbf{R}$ be integrable functions such that

$$\tilde{\mathcal{E}} = \tilde{\mathcal{E}}_1 + \ldots + \tilde{\mathcal{E}}_n. \tag{40}$$

Now suppose that for all $i \in \{1, \ldots, n\}$ we have

$$\int_{\theta_1}^{\theta_2} \tilde{\mathcal{E}}_i(\psi)\, d\psi = \int_{u_{i-1}}^{u_i} \mathcal{I}(s)\, ds = u_i - u_{i-1}, \tag{41}$$

then each part U_i of the reflector can realize the contribution $\tilde{\mathcal{E}}_i$ to the total required distribution. So, for fixed n, U_i and $\tilde{\mathcal{E}}_i$, we may consider solutions θ such that θ is smooth (and thus monotonic) on each open interval (u_{i-1}, u_i) and such that U_i realizes the contribution $\tilde{\mathcal{E}}_i$ to $\tilde{\mathcal{E}}$. Since on each interval, we can choose either the increasing or the decreasing solution, this generally leads to 2^n possible solutions for this particular choice of the intervals U_i and

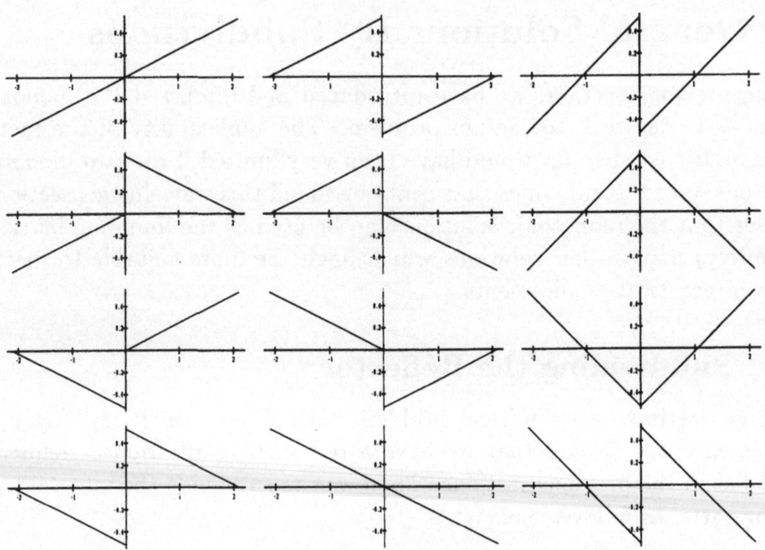

Figure 6: The 12 θ's corresponding to the disjoint (first column), the crossing (second column) and the overlapping (third column) subdivisions.

the function $\tilde{\mathcal{E}}_i$. Needless to say, there are infinitely many ways of choosing these.

In order to investigate subdivisions in more detail, let us consider subdivisions into 2 parts only. (These can be used repeatedly.) In order to subdivide the reflector, we must specify the subdivision direction $u_1 \in [t_1, t_2]$. The functions $\tilde{\mathcal{E}}_1$ and $\tilde{\mathcal{E}}_2$ should then be chosen such that (41) holds. Again, there are infinitely many possible choices, but there are some straightforward ones which are easily computed and practically relevant. These will be described below and illustrated by the corresponding solutions to Example 3.1.

(i) The *disjoint* subdivision. In this subdivision, we let the lower and upper part of the reflector realize the lower and upper part of the required distribution, respectively. To that end, we have to find a direction σ such that

$$\int_{\theta_1}^{\sigma} \tilde{\mathcal{E}}(\psi)\, d\psi = u_1 - t_1. \tag{42}$$

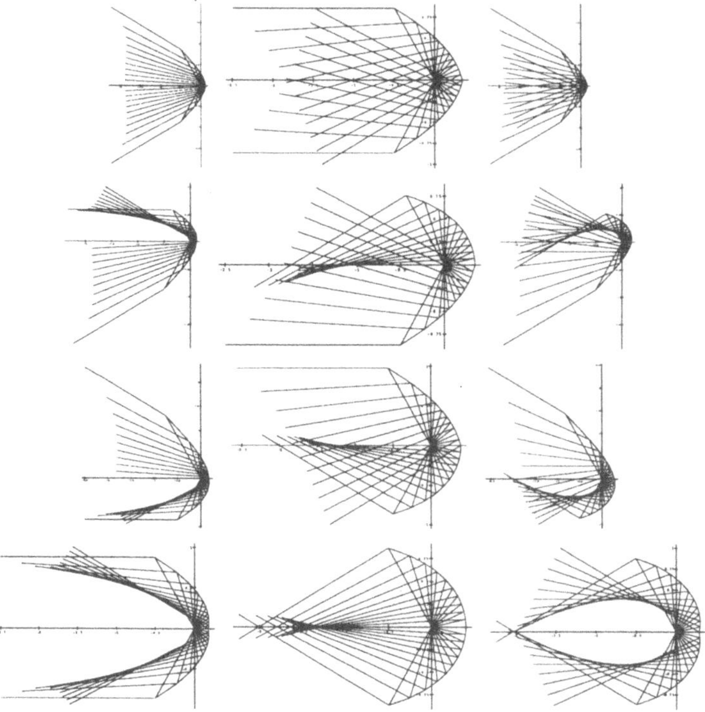

Figure 7: The 12 reflectors corresponding to the disjoint (first column), the crossing (second column) and the overlapping (third column) subdivisions.

Then we can set

$$
\tilde{\mathcal{E}}_1(\psi) = \begin{cases} \tilde{\mathcal{E}}(\psi) & \text{if } \psi < \sigma, \\ 0 & \text{otherwise,} \end{cases} \tag{43}
$$

$$
\tilde{\mathcal{E}}_2(\psi) = \begin{cases} 0 & \text{if } \psi < \sigma, \\ \tilde{\mathcal{E}}(\psi) & \text{otherwise.} \end{cases} \tag{44}
$$

(ii) The *crossing* subdivision. In this subdivision, we let the lower and upper part of the reflector realize the upper and lower part of the required distribution, respectively. Again, the two parts of the reflector realize disjoint parts of the required distribution, but this time the two ray bundles cross.

To subdivide the intensity distribution correspondingly, we have to find a direction τ such that

$$\int_{\theta_1}^{\tau} \tilde{\mathcal{E}}(\psi)\, d\psi = t_2 - u_1. \tag{45}$$

Then we can set

$$\tilde{\mathcal{E}}_1(\psi) = \begin{cases} 0 & \text{if } \psi < \tau, \\ \tilde{\mathcal{E}}(\psi) & \text{otherwise,} \end{cases} \tag{46}$$

$$\tilde{\mathcal{E}}_2(\psi) = \begin{cases} \tilde{\mathcal{E}}(\psi) & \text{if } \psi < \tau, \\ 0 & \text{otherwise.} \end{cases} \tag{47}$$

(iii) The *overlapping* subdivision. In this subdivision, we let the lower and upper part of the reflector both take care of proportional parts of $\tilde{\mathcal{E}}$, as follows. Let

$$\tilde{\mathcal{E}}_1(\psi) = \frac{u_1 - t_1}{t_2 - t_1} \tilde{\mathcal{E}}(\psi), \tag{48}$$

$$\tilde{\mathcal{E}}_2(\psi) = \frac{t_2 - u_1}{t_2 - t_1} \tilde{\mathcal{E}}(\psi). \tag{49}$$

When we apply all these subdivisions to Example 3.1 with $u_1 = 0$, we obtain the 12 corresponding functions θ and reflectors as shown in Figs. 6 and 7. In these figures, the disjoint, crossing and overlapping solutions are shown in the first, second and third columns, respectively. In the first row, both lower and upper parts yield increasing raypaths; in the second row, the lower and upper parts yield increasing and decreasing raypaths, etc.

We see that the two monotonic solutions are present here as well, and that there are two more smooth solutions corresponding to the symmetric overlapping subdivisions. Note furthermore that the shapes of these reflectors, which all have a fixed lower endpoint $v(t_1)$, differ significantly. In Fig. 8 all 12 reflectors are drawn.

4.2   Discussion on Solutions by Subdivisions

By choosing an appropriate subdivision, an experienced designer can try to find a reflector that produces the required distribution by a ray path that suits additional constraints. For instance, in two solutions in Figure 7, we see that no reflected ray goes through the origin. Such solutions may

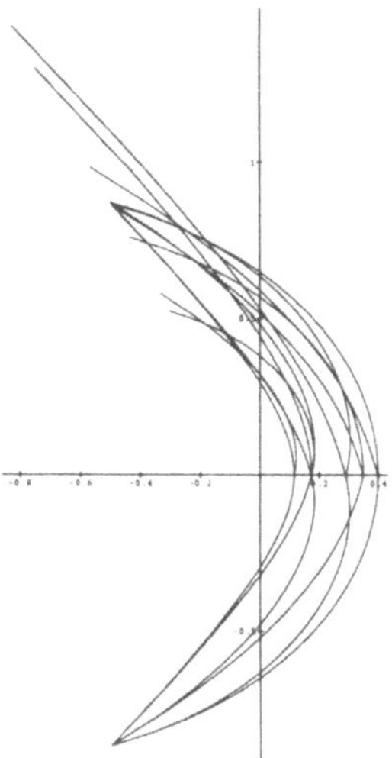

Figure 8: The 12 solutions to Example 3.1.

be important when one wishes to take the finite dimensions of real light sources into account and when one wants to avoid that reflected rays reenter the source. Also, in near field problems, the angle of incidence of rays on a screen may be important. Different subdivision solutions will also have different properties concerning the angles of incidence.

This way of dealing with side constraints not only increases the practical applicability of these mathematical methods, it also introduces some challenging new mathematical problems. One may ask which solutions may cause multiple reflections and which not; which solutions are more robust to inaccurate lamp positioning. In [7] and [4], it has been investigated how the choice of the subdivision affects the size of the reflector.

In order to really exploit the existance of multiple solutions to meet other constraints, we either have to have very experienced designers, or we need more mathematical research into these directions.

4.3 Extensions to 3-Dimensional Problems

In the previous sections we have seen that mathematical methods for 2-dimensional reflector design should contain much more than just solving a differential equation, before real practical applicability is possible.

When we consider the 3-dimensional problem for a point source, a lot of research has been done over the past 20 years. Schruben [12] first introduced the basic partial differential equation that describes the solutions for 3-dimensional problems. The basic equation is of Monge-Ampere type: a reflector, described by a function $r(\theta, \phi)$ (where θ and ϕ represent polar coordinates) is given by the equation

$$A(r_{\theta\theta}r_{\phi\phi} - r_{\theta\phi}^2) + Br_{\theta\theta} + Cr_{\theta\phi} + Dr_{\phi\phi} + E = 0, \qquad (50)$$

where A, B, C, D and E are functions of $\theta, \phi, r, r_\theta$ and r_ϕ.

Several studies on this equation related to the reflector design problem have been carried out by Westcott et al.; see e.g. [15, 2, 9, 14, 1, 3]. Existance and uniqueness results have for some special cases been given by Marder [8] and Oliker [10, 11].

While an analytical approach based on this equation has been successfully applied to e.g. reflector antenna design, it is unlikely that this approach will soon be applicable to other reflector design problems in lighting industry, where all sorts of practical constraints have to be dealt with. Even if we can solve (50), we have only made a small start.

References

[1] F. Brickell, L. Marder and B.S. Westcott, The geometrical optics design of reflectors using complex coordinates, *J. Phys. A: Math. Gen.* **10**, No. 2, 1977, pp. 245-260.

[2] F. Brickell and B.S. Westcott, Reflector design for two-variable beam shaping in the hyperbolic case, *J. Phys. A: Math. Gen.* **9**, No. 1, 1976, pp. 113-128.

[3] F. Brickell and B.S. Westcott, Phase and power density distributions on plane apertures of reflector antennas, *J. Phys. A: Math. Gen.* **11**, No. 4, 1978, pp. 777-789.

[4] A.J.E.M. Janssen and M.J.J.J.B. Maes, An optimization problem in reflector design, *Philips Journal of Research*, **47**, 1992, pp. 99-143.

[5] H.A.E. Keitz, *Light Calculations and Measurements*, MacMillan and Co. Ltd., London, 1971.

[6] J.B. Keller, The inverse scattering problem in geometrical optics and the design of reflectors, *IRE Trans. Antenn. Propag.*, 1958, pp. 146-149.

[7] M.J.J.J.B. Maes and A.J.E.M. Janssen, A note on cylindrical reflector design, *Optik* **88**, No. 4, 1991, pp. 177-181.

[8] L. Marder, Uniqueness in reflector mappings and the Monge-Ampère equation, *Proc. R. Soc. Lond.* A **378**, 1981, pp. 529-537.

[9] A.P. Norris and B.S. Westcott, Computation of reflector surfaces for bivariate beamshaping in the elliptic case, *J. Phys. A: Math. Gen.* **9**, No. 12, 1976, pp. 2159-2169.

[10] V.I. Oliker, Near radially symmetric solutions of an inverse problem in geometric optics, *Inverse Problems* **3**, 1987, pp. 743-756.

[11] V.I. Oliker, On reconstructing a reflecting surface from the scattering data in the geometric optics approximation, *Inverse Problems* **5**, 1989, pp. 51-65.

[12] J.S. Schruben, Formulation of a reflector-design problem for a lighting fixture, *J. Opt. Soc.* **62**, No. 12, 1972, pp. 1498-1501.

[13] V.V. Trembač: *Luminaires*. Soviet Union Publishing Institute for Energy Technology, Moscow, Leningrad 1958.

[14] B.S. Westcott and F. Brickell, Computation of reflector surfaces for two-variable beam shaping in the hyperbolic case, *J. Phys. A: Math. Gen.* **9**, No. 4, 1976, pp. 611-625.

[15] B.S. Westcott and A.P. Norris, Reflector synthesis for generalized far-fields, *J. Phys. A: Math. Gen.* **8**, No. 4, 1975, pp. 521-532.

RECONSTRUCTION OF MULTIPLE CRACKS FROM EXPERIMENTAL, ELECTROSTATIC BOUNDARY MEASUREMENTS.

Kurt Bryan, Valdis Liepa, and Michael Vogelius.

1 Introduction

The purpose of this presentation is to demonstrate the viability of using Electrical Impedance Tomography (EIT) for the reconstruction of multiple macroscopic cracks. Through the use of EIT one seeks to determine the interior conductivity properties of a specimen from measurements of electrostatic potentials and currents on the boundary. There are obvious applications of this technology to many problems, for instance in medical tomography [5, 8] or nondestructrive testing of mechanical components [10]. There are two very separate classes of practical reconstruction techniques: 1) general imaging techniques, and 2) techniques that use prior information or assumptions. The first approach produces a somewhat blurry image of the conductivity distribution inside the specimen (somewhat akin to an ultrasound-scan) from which an "educated" user may then draw conclusions of a more specific nature (*cf.* [5, 8, 10, 18]). The second approach uses a restricted model for the interior conductivity distribution and seeks to determine more specific features within this model which are consistent with the measured data (*cf.* [13, 15, 16]). This may be done probabilistically, in a way reminiscent of methods for image reconstruction known as "maximum entropy methods" or deterministically by seeking to fit a relatively small number of parameters. In this presentation we shall concentrate on a very specific method of this last category. It seeks to determine a finite number of linear cracks consistent with the boundary measurements. In particular we shall test this algorithm on data which has been collected from laboratory experiments.

The mathematical results which insure that a finite number of cracks can be reconstructed from a finite set of boundary voltages and boundary currents are found in [6, 11]. In these papers it is proven that n C^2 cracks inside a two dimensional domain may be determined from $n + 1$ pairs of boundary voltages and boundary currents. These results have recently been extended to show that measurement of two pairs of boundary voltages and boundary currents suffice to determine any number of cracks [2, 9]. It is very easy to see that one measurement will not suffice to determine even a single crack [11].

The computational algorithm we shall employ has largely been developed in [7]. It is the natural extension of an algorithm which was developed in [17] for the case of a single crack. We have modified the algorithm from [7] in one important respect, by making the selection of the initial guess automatic. For this purpose we have implemented some formulae which were suggested in [4] and which directly reconstruct the line on which a single linear crack must lie in order to be consistent with a single pair of boundary voltages and boundary currents. These formulae may occasionally be degenerate, but the use of additional measurement pairs can eliminate this degeneracy.

An outline of this presentation is as follows. In section 2 we introduce the mathematical model used for simulating cracks and we give a brief discussion of our computational reconstruction method. In section 3 we present the formulae derived by Andrieux and Ben Abda and we briefly describe the selection of the initial crack. Section 4 contains a description of the experimental equipment used to collect the actual data for our reconstructions, and in particular we describe the few modifications that were made when compared to the equipment used earlier to collect data for the reconstruction of a single crack [17]. In the fifth and final section we provide a number of examples of reconstructions of multiple cracks based on the collected experimental data.

2 The Mathematical Model and the Reconstruction Algorithm

A single crack inside a two dimensional conductor is commonly modeled as a perfectly insulating curve σ. With a background conductivity $0 < \gamma_0 \leq \gamma(x) \leq \gamma_1$ and a finite collection of cracks $\Sigma = \cup_{k=1}^{n} \sigma_k$, the steady state conductance equations thus read

$$\nabla \cdot (\gamma \nabla v) \; = \; 0 \quad \text{in} \quad \Omega \setminus \Sigma, \tag{2.1}$$

$$\gamma \frac{\partial v}{\partial \nu} \; = \; 0 \quad \text{on} \quad \Sigma,$$

with appropriate boundary conditions on $\partial \Omega$, $e.g.$,

$$v = \phi \quad \text{on} \quad \partial \Omega. \tag{2.2}$$

The field ν is normal to Σ. The function v represents the potential induced in Ω. We assume that $\Omega \subset I\!R^2$ is simply connected, $i.e.$, has no holes, and so the entire boundary $\partial \Omega$ is accessible from the "outside". The domain Ω corresponding to our experimental data is a disk. Let u denote the "γ-harmonic" conjugate to v. It is related to v by the formula

$$(\nabla u)^{\perp} = \gamma \nabla v \quad \text{in} \quad \Omega \setminus \Sigma, \tag{2.3}$$

where $\perp$ indicates counterclockwise rotation by $\pi/2$. Note that the set $\Omega \setminus \Sigma$ is not simply connected; the existence of a "γ-harmonic" conjugate u owes not only to the fact that $\nabla \cdot (\gamma \nabla v) = 0$ in $\Omega \setminus \Sigma$, but also to the fact that $\gamma \frac{\partial v}{\partial \nu} = 0$ on Σ. For a particular set of constants c_k, $k = 1, \dots, n$, the function u solves the problem

$$\nabla \cdot (\gamma^{-1} \nabla u) \; = \; 0 \quad \text{in} \quad \Omega \setminus \Sigma, \tag{2.4}$$

$$u \; = \; c_k \quad \text{on} \quad \sigma_k, \; k = 1, \dots, n$$

with

$$\gamma^{-1} \frac{\partial u}{\partial \nu} = \psi = \frac{\partial \phi}{\partial s} \quad \text{on} \quad \partial \Omega. \tag{2.5}$$

Here s denotes the counterclockwise tangent direction on $\partial \Omega$ and ν denotes the outward normal on $\partial \Omega$. For these particular constants, finding a solution to (2.4)-(2.5) is thus equivalent to finding a solution to (2.1)-(2.2). The constants c_k may (up to a common additive constant) be characterized in several equivalent ways (see [7]). The characterization we shall use here takes the form of the system of n equations

$$\int_{\sigma_k} [\gamma^{-1} \frac{\partial u}{\partial \nu}] \, ds = 0, \quad 1 \leq k \leq n. \tag{2.6}$$

Here $[\gamma^{-1}\frac{\partial u}{\partial \nu}] = \gamma^{-1}\frac{\partial u}{\partial \nu_+} - \gamma^{-1}\frac{\partial u}{\partial \nu_-}$ denotes the jump in the normal flux across the curve σ_k. [1] The system (2.6) should be viewed as a supplement to the boundary value problem (2.4)-(2.5). The solution u has a physical interpretation in its own right – it is the potential generated by the boundary current ψ for a medium with background conductivity γ^{-1} and a set of perfectly conducting cracks σ_k, $k = 1, \ldots, n$. As follows immediately from the relation (2.3), one passes between the boundary data corresponding to v and that corresponding to u by differentiating the Dirichlet-data along the boundary, integrating the Neumann-data along the boundary and interchanging the roles of the two. Except when we explicitly say so we shall always work with cracks that are perfectly conducting in the sense described above. Let $P_1, \ldots, P_M$ and $Q_1, \ldots, Q_M$ be $2M$ points on $\partial\Omega$. For the crack reconstruction we utilize solutions corresponding to the two-electrode currents $\psi_j = \delta_{P_j} - \delta_{Q_j}$, $j = 1, \ldots, M$,

$$\nabla \cdot (\gamma^{-1}\nabla u_j) \;=\; 0 \quad \text{in} \quad \Omega \setminus \Sigma, \tag{2.7}$$
$$u_j \;=\; c_k^{(j)} \quad \text{on} \quad \sigma_k, \; k = 1, \ldots, n,$$

with

$$\gamma^{-1}\frac{\partial u_j}{\partial \nu} = \delta_{P_j} - \delta_{Q_j} \quad \text{on} \quad \partial\Omega, \tag{2.8}$$

the constants $c_k^{(j)}$ being determined through the additional equations (2.6). The reconstruction problem may now be stated explicitly as follows:

We seek to reconstruct the collection of cracks $\Sigma = \cup_{k=1}^n \sigma_k$ *from knowledge of the boundary voltage data* $\{u_j|_{\partial\Omega}\}_{j=1}^M$ *corresponding to the prescribed two-electrode currents* $\gamma^{-1}\frac{\partial u_j}{\partial \nu} = \delta_{P_j} - \delta_{Q_j}$, $j = 1, \ldots, M$.

It was shown in [6] that if we take $M = n + 1$, $Q_1 = Q_2 \ldots = Q_{n+1} = P_0$ and we take P_j, $1 \leq j \leq n + 1$ to be mutually different and different from P_0, then the boundary voltage measurements corresponding to the resulting $n + 1$ fixed two-electrode currents $\delta_{P_j} - \delta_{P_0}$, $1 \leq j \leq n + 1$ suffice to uniquely identify a collection of n (or fewer) cracks. By a clever extension of this analysis it has recently been shown that two fixed two-electrode currents suffice to uniquely identify any number of cracks, [2, 9]. It is very easy to see (*cf.* [11]) that measurements corresponding to a single boundary current do not suffice to determine even a single crack. In the papers [6, 11] it was required that the reference conductivity γ be real analytic. This requirement has recently been considerably relaxed; the identifiability result holds even if γ is only Hölder continuous [9].

A certain amount of knowledge is available concerning continuous dependence. In the case in which the background conductivity is constant and there is only one crack present it has been shown (*cf.* [1]) that if the boundary voltage data (on some open subset of $\partial\Omega$) deviate by ϵ, then the crack locations differ by at most $C[\log(|\log \epsilon|)]^{-1/4}$. If the single crack is linear it has been shown that the crack location depends Lipschitz continuously on the boundary data, [3]. This last result significantly extends a calculation found in [11], showing that transverse translations and rotations of a single linear crack depend Lipschitz continuously on the measured data.

As mentioned before, we shall in the present reconstructions always try to fit the data by means of linear cracks. We shall also assume that the background conductivity may be modeled by a constant, *e.g.*, $\gamma \equiv 1$. As was the case in [17], we base the reconstruction of the cracks on the values

---

[1]The expression $\frac{\partial u}{\partial \nu_+}$ denotes the limit of the derivative (in the direction ν) as one approaches σ_k from the side to which ν points. $\frac{\partial u}{\partial \nu_-}$ denotes the limit as one approaches σ_k from the opposite side.

of a relatively small number of functionals (as opposed to all the boundary measurements). In [17] we used 4 functionals for the reconstruction of a single linear crack, here we use $4n$ functionals for the reconstruction of n cracks. We now give a brief description of these functionals. For more details we refer the reader to [7, 17].

Let $\mathbf{F}$ denote the vector-valued function

$$\mathbf{F}(\Sigma, \psi, \mathbf{w}) = (F(\Sigma, \psi, w^{(1)}), F(\Sigma, \psi, w^{(2)}), F(\Sigma, \psi, w^{(3)}), F(\Sigma, \psi, w^{(4)}))^t,$$

where $F(\Sigma, \psi, w)$ is given by

$$F(\Sigma, \psi, w) = \int_{\partial\Omega} u(\Sigma, \psi)\frac{\partial w}{\partial \nu}\, ds. \tag{2.9}$$

The functions $w^{(i)}$, $1 \leq i \leq 4$, will be taken as particular solutions of

$$\triangle w = 0 \quad \text{in} \quad I\!\!R^2 \setminus \Sigma.$$

The function $u = u(\Sigma, \psi)$ is the solution to (2.4)-(2.5) with $\gamma \equiv 1$. In order to make u unique we may for instance impose the requirement

$$\int_{\partial\Omega} u\, ds = 0 \ . \tag{2.10}$$

The appropriate selection of boundary currents ψ and test functions $\mathbf{w} = (w^{(1)}, w^{(2)}, w^{(3)}, w^{(4)})^t$ is very important and was discussed in detail in [7]. For our reconstruction algorithm we always choose ψ in the form of a two-electrode current. We take one ψ and one $\mathbf{w}$ corresponding to each crack σ_k; whenever we want to emphasize this correspondence we use the notation

$$\psi_{\sigma_k} \quad \text{and} \quad \mathbf{w}_{\sigma_k} = (w^{(1)}_{\sigma_k}, w^{(2)}_{\sigma_k}, w^{(3)}_{\sigma_k}, w^{(4)}_{\sigma_k})^t.$$

We shall always take $\mathbf{w}$ so that

$$\int_{\partial\Omega} \frac{\partial w^{(i)}}{\partial \nu}\, ds = 0, \quad \text{and} \quad \int_\sigma \left[\frac{\partial w^{(i)}}{\partial \nu}\right]\, ds = 0 \ \ \forall \sigma \in \Sigma. \tag{2.11}$$

Because of the first identity in (2.11) the function $\mathbf{F}$ is unchanged by the addition of a constant to u, and we can therefore work with any other normalization in place of (2.10). The data for our reconstruction consist of measured boundary potentials. In practice we can of course only measure the values of these potentials at a finite (fairly moderate) number of points. We denote by $g(\psi)$ the voltage data corresponding to the boundary current ψ, and we define a corresponding vector-valued function

$$\mathbf{f}(\psi, \mathbf{w}) = (f(\psi, w^{(1)}), f(\psi, w^{(2)}), f(\psi, w^{(3)}), f(\psi, w^{(4)}))^t,$$

where $f(\psi, w)$ is given by

$$f(\psi, w) = \int_{\partial\Omega} g(\psi)\frac{\partial w}{\partial \nu}\, ds, \tag{2.12}$$

and where $w^{(i)}$ are the same functions as before. Our algorithm seeks a solution $\Sigma = \{\sigma_k\}_{k=1}^n$ to the $4n$ equations

$$\mathbf{F}(\Sigma, \psi_{\sigma_k}, \mathbf{w}_{\sigma_k}) = \mathbf{f}(\psi_{\sigma_k}, \mathbf{w}_{\sigma_k}), \quad 1 \leq k \leq n. \tag{2.13}$$

We do not use information about the full set of measured boundary voltages for the reconstruction; we only use information about the values of these particular functionals. This is in contrast to the reconstructions from experimental data which we presented in [13], and where we used all the measured boundary voltages and a least squares approach. Based on extensive experimentation it is our experience that almost all the relevant information is contained in these functionals. Admission of the extra data, *e.g.* after convergence of the algorithm based on the functionals, does not generally seem to improve the reconstructions. In several examples in which we tried the least squares approach from the start it actually impeded the convergence process; we think this phenomenon may be caused by the presence of local minima.

The construction of the test functions $w^{(i)}$ is quite simple to describe. If we consider a single linear crack, and use a coordinate system so that the crack lies on the x-axis (the line of reals) then the functions $w^{(1)}$ and $w^{(2)}$ are given by

$$w^{(1)} = Im[z], \quad w^{(2)} = Im[z^2] \ .$$

The functions $w^{(3)}$ and $w^{(4)}$ have square-root singularities at the endpoints of the crack; if the crack lies between the origin and the point $(\lambda, 0)$ they have the form

$$w^{(3)} = \begin{cases} Re[(z - \lambda)\sqrt{z(z - \lambda)}], & Re(z) > \frac{\lambda}{2} \\ -Re[(z - \lambda)\sqrt{z(z - \lambda)}], & Re(z) < \frac{\lambda}{2} \ , \end{cases} \tag{2.14}$$

$$w^{(4)} = \begin{cases} Re[\sqrt{z(z - \lambda)}], & Re(z) > \frac{\lambda}{2} \\ -Re[\sqrt{z(z - \lambda)}], & Re(z) < \frac{\lambda}{2} \ . \end{cases} \tag{2.15}$$

Intuitively $w^{(1)}$ and $w^{(2)}$ detect transverse translations and rotations of the crack, whereas $w^{(3)}$ and $w^{(4)}$ detect translations along the direction of the crack and variations in length. It is interesting to note that in the limit as λ approaches zero the information extracted by these four functions approach the first four nontrivial Fourier modes of the boundary data. We refer the reader to [7] for more details.

The selection of the applied currents is also fairly simple to describe: for a single crack we select the location of the two electrodes in a way which maximizes the sensitivity of the four functionals with respect to rotations and transverse translations of the crack. We use an iterative approach to solve (2.13), in this case for the single crack σ. At each step new electrode locations are found by maximizing two of the diagonal elements of the Jacobian of $D_\sigma F$ with respect to the boundary electrode locations (taking the fixed test functions described above). Measured data corresponding to these electrode locations are then used to find an updated crack. The selection procedure is described in detail in [7, 17]. When the domain is a disk, and the crack is of moderate size and sits away from the boundary, then the most sensitive electrode locations defined above turn out (very nearly) to be the points that arise as the intersection of the line on which the crack lies and the domain boundary (remember, the crack is perfectly conducting). Our selection of electrodes does require the solution of an auxiliary boundary value problem. However, as explained later, this is done very efficiently through an integral equation formulation. For the reconstruction of more than one crack we select a boundary current corresponding to each crack, as if the other cracks were not present. This process is described in detail in [7].

The $4n \times 4n$ system (2.13) is solved by means of a Newton method. As explained above, the test functions and the applied currents change with each step of the iteration. The update for each step of the Newton method is constrained, along the lines of the classical Levenberg-Marquardt approach [14].

The boundary value problems which must necessarily be solved in order to evaluate F (and find the "optimally sensitive" electrode locations) are formulated in terms of boundary integral equations. The system of boundary integral equations is a mixture of first and second kind equations. There is a first kind contribution corresponding to the functions that describe the normal derivative "jumps" on each crack. However, by working with the boundary condition $u =$constant on each crack, we avoid the problem associated with non-integrable kernels. We discretize the integral equations by Nyström's method, as completely described in [7]. The kernels of the first kind contributions are singular enough that we did not find it necessary to regularize the equations. Also, we note that we do not really numerically compute the $u|_{\partial\Omega}$ necessary to evaluate F, rather we compute $(u - u_0)|_{\partial\Omega}$, where u_0 is the potential generated by the exact same electrode locations and currents as u, but without any cracks present. The computation of the smooth function $(u - u_0)|_{\partial\Omega}$ is a much better posed problem than the computation of $u|_{\partial\Omega}$. Because the domain Ω is a disk, the function u_0 has a very simple explicit expression. Whenever necessary we add its values to the computed differences.

3   Apparatus and Data Collection

In this section we give a brief description of the apparatus with which we collected the experimental data used to test the crack reconstruction algorithm. Most of these details can also be found in [13]; nonetheless, for the convenience of the reader we will recount them here.

The reconstruction algorithm is designed for a two dimensional region with a constant background conductivity. An experimental equivalent is given by the tank illustrated in Figure 1. The tank is cylindrical and constructed of 0.25 inch thick plexiglass, with an inside diameter of 10.5 inches and a height of 23.5 inches. The tank is filled with distilled water to which a small amount of ordinary tap water is added, which gives the solution a modest electrical conductivity. Twelve evenly spaced copper electrodes run vertically along the full height of the tank. The copper electrodes are 0.25 inches wide. Since the entire apparatus is uniform in the vertical direction, and since the electrodes are essentially perfectly conducting (in relation to the interior solution), the tank provides a reasonable approximation to a two-dimensional electrical conduction problem. To introduce perfectly conducting "cracks" we insert strips of sheet metal with widths ranging from 2 to 6 inches vertically into the tank, again preserving the uniformity of the conduction problem in the vertical direction. A grid is placed under the tank (which has a plexiglass bottom) so that the location of the cracks can be easily recorded.

The overall structure of the data collection apparatus is as follows. Each electrode is attached to a binding post on the top rim of the tank. These binding posts are connected, through a bank of relay cards, to a voltage source and a pair of digital multimeters. One multimeter serves to measure the electrical current supplied to the tank when a specified pair of electrodes is active. The other multimeter measures the induced voltage drop between any given pair of electrodes. By altering the electrical connections via the relay cards, any pair of electrodes can act as the source

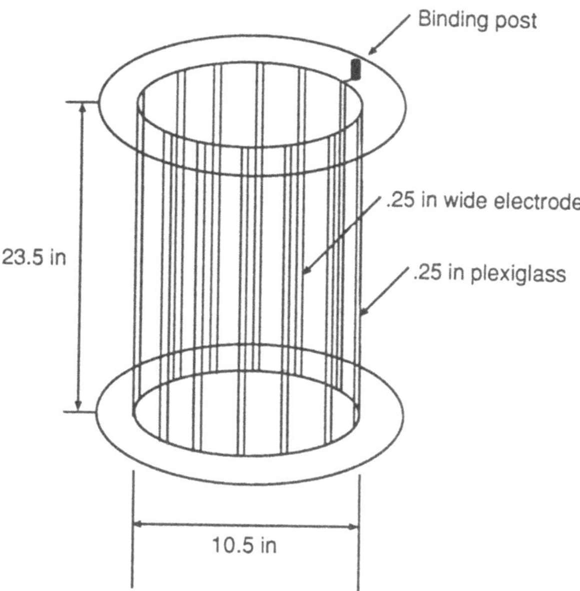

Figure 1: Tank used for data collection.

and sink for the electrical current and the voltmeter can measure the resulting potential across any two electrodes. The entire apparatus is connected to a PC which controls the relay cards, allowing software switching of the active electrodes and voltage measurements. The PC and software also communicate with the multimeters, collect the resulting current and voltage information, and present it in a convenient format. A simple schematic of the entire apparatus is given in Figure 2.

As mentioned in [13], while the problem is modeled as a steady-state or DC conduction problem, in practice one uses an AC current. This is to avoid any electrochemical plating of the electrodes with the impurities present in the water, which might alter the behavior of the electrodes, particularly the contact resistance between the electrodes and water. We use a sinusoidal source with a frequency of 1 kHz, which should be low enough to be considered steady-state so that no phase changing capacitance or inductance effects will be significant. In actuality, a few experiments were conducted with frequencies from 10 Hz to 10 kHz without any cracks in the tank, and in this range no frequency dependent effects were observed.

The data collection software is written in BASIC. As discussed in the previous section, at the heart of the crack recovery algorithm is the idea of iteratively adapting the active electrode locations for greater sensitivity as the algorithm progresses and more precise information about the actual crack location(s) becomes available. However, since the reconstruction algorithm could not

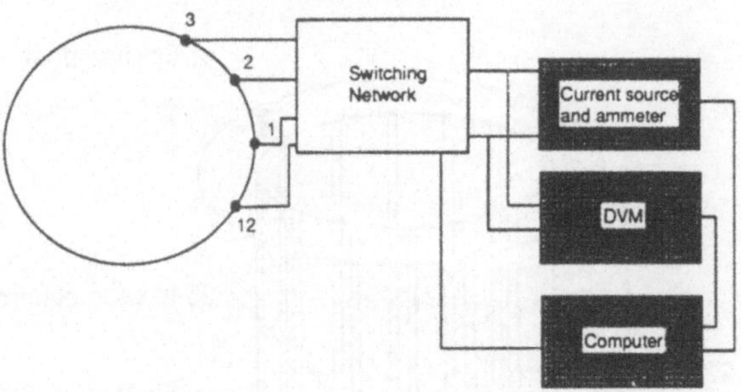

Figure 2: Schematic of data collection system.

immediately be ported to the PC which collects the data, it was necessary to collect all possible data a priori. When the reconstruction algorithm was later run using this data set (on a separate computer), it simply extracted the required voltage data corresponding to the desired active electrode locations. One can easily check that for a tank with 12 electrodes there are only 11 different linearly independent combinations of input and output electrodes; the solution corresponding to any other current pattern can be represented as a linear combination of the solutions corresponding to these basic current patterns. In its dipole current pattern mode the software collects data for the current patterns $\psi_j = \delta_{j+1} - \delta_j$, $j = 1, \ldots, 11$, that is, a current is introduced at electrode $j+1$ and withdrawn at electrode j. Whenever the recovery algorithm needs a different current pattern during its execution it simply takes an appropriate linear combination of the data corresponding to these 11 basic patterns.

For the data sets which will be presented, typical values for the input current were 4 to 7 mA (rms). Typical voltage drops between electrodes range from zero to 7 volts rms, with the largest drops between the active input and output electrode. The previous version of the software, used to collect the data in [13], recorded the input current to only 2 significant figures. In the course of taking measurements for this paper it was observed that even the third significant figure for the current was quite consistent and stable from measurement to measurement. The software was thus modified to record three significant figures for the input current. As before, voltages are recorded to four figures. An application of the crack reconstruction algorithm to both types of data (two versus three significant figures for current measurement) indicate that the latter is considerably more accurate. It appears that for this experimental setup, the crack manifests much information concerning its presence and location in this third significant digit, particularly for multiple crack problems.

Before proceeding with the reconstruction algorithm one must also know k, the background conductivity of the tank. To obtain this information we applied the current patterns $\psi_1, \ldots, \psi_{11}$ described above to the tank with no cracks and measured the corresponding induced voltages. We then used the same current patterns in our computational model and computed the corresponding theoretical voltage. These voltages scale linearly with $\frac{1}{k}$, and so one can easily compute that value for k which gives the best fit of the computational model to the measured voltages, in the sense of mean square error. Doing this yielded an average resistivity of 456.6 ohms, or a conductivity of 0.00219 mhos. This value varied by only 1% over the course of the tests (approximately 30 different crack configurations over 2 days). This is the value of k which was used in all crack reconstructions. In practice we scaled the measured data by k so that it formally corresponds to a background conductivity of one. All our subsequent computations were then done with a conductivity of 1.

One final remark is in order. The electrodes on the actual tank have a necessarily non-zero width, while the simple theoretical model we use assumes point electrodes. This assumption leads to logarithmic singularities in the theoretical voltage at the current input and output electrodes. Moreover, in the actual experimental data any contact resistance between the electrode surface and the water leads to inaccuracies in the voltage measurements at the active electrodes, since Ohm's law implies that if a current is flowing through the electrode and a resistance is present, there must be a voltage drop across the electrode. Thus the voltage on the tank side of the active electrode (the voltage required by the algorithm) may differ from the voltage measured by the multimeter which is on the other "side" of the contact resistance. This is not a problem for voltage measurements at non-active electrodes; these electrodes have essentially no current flowing (the multimeter presents an input impedance of $> 10^9$ ohms), hence no voltage drop. We thus omit from consideration the voltages at active electrodes for both the experimental and computational data. One merely adjusts the numerical integration rule for computing the functionals $F(\Sigma, \psi, w^{(i)})$ and $f(\psi, w^{(i)})$ to account for the gap.

While a more accurate model incorporating a non-zero width electrode could certainly be derived and the algorithm adapted to deal with such a model, the agreement between the measured data and theoretical computations was so good that there seemed little to be gained by refining the model, at least for cracks which are not extremely close to an active electrode.

4　A Simple Formula for the Line of a Single Crack

In [4] two very simple formulae are given for reconstructing the line on which a single linear, perfectly insulating crack lies. These formulae are in terms of a single voltage/current pair measured on the boundary of Ω. The formulae are restricted to the reconstruction of a single line. As has already been observed, a single crack (even a linear one) is not uniquely identifiable from a single measurement. As far as these explicit formulae are concerned this expresses itself in terms of the possibility of the vanishing of a certain denominator (more about this later). The formulae do not require the solution of any boundary value problems, but only the evaluation of certain weighted integrals of the voltage and current data on the boundary of the region. These formulae merely locate the line on which the crack lies, they do not tell us where on the line the crack is. It is possible to determine the exact location of the crack by evaluating an infinite number of similar integrals and reconstructing the support of a function from its Fourier series (see [4]). We have not implemented this latter step as part of our algorithm, but we have incorporated the simple formulae for the reconstruction of a single line as part of the initialization step of our crack reconstruction algorithm. Once the line is found, the initial guess is chosen to be centered on that segment of the line which lies inside the domain (a disk). In the reconstructions we present later the initial guess is, rather arbitrarily, taken to be of length $0.1\times$ the radius of the disk.

We now briefly review the formulae from [4]. Suppose that σ is a perfectly insulating linear crack in Ω and $v(x)$ is a function on $\Omega \setminus \Sigma \subset I\!\!R^2$ which satisfies

$$\Delta v = 0 \quad \text{in} \quad \Omega \setminus \sigma$$
$$\frac{\partial v}{\partial \nu} = 0 \quad \text{on} \quad \sigma ,$$

where ν is a unit normal vector field on σ. Note that this field is constant ($\nu = \nu_0$) since σ is linear. Based on the boundary values of v and $\frac{\partial v}{\partial \nu}$ on $\partial \Omega$ we can then recover the line $\nu_0 \cdot x = c$ on which the crack σ lies as follows. Let $\phi_i(x)$ denote any of the two coordinate functions $\phi_i(x) = x_i$. Note that ϕ_i are both harmonic. Define the quantities I_i, $i = 1, 2$, by

$$I_i = \int_{\partial \Omega} \left(v \frac{\partial \phi_i}{\partial \nu} - \phi_i \frac{\partial v}{\partial \nu} \right) ds, \quad \phi_i(x) = x_i. \tag{4.16}$$

These quantities depend only on boundary values of v and its normal derivative, and can be easily computed from the known data, e.g., when $\frac{\partial v}{\partial \nu}$ is specified (a current applied) and v is measured on $\partial \Omega$, or vice-versa. Let $[v]$ denote the jump in u across σ; $[v]$ is defined as $v_+ - v_-$, where v_+ denotes the values of v as one approaches σ from the side to which the normal vector ν_0 points, and v_- denotes the values as one approaches from the opposite side. A simple application of the divergence theorem shows that

$$I_i = \int_{\sigma} [v] \frac{\partial \phi_i}{\partial \nu} ds ,$$

since $\frac{\partial v}{\partial \nu}$ vanishes on σ. Now note that $\frac{\partial \phi_i}{\partial \nu} \equiv \nu_{0,i}$ on σ, since $\phi_i = x_i$. Thus

$$I_i = \nu_{0,i} \int_{\sigma} [v] ds . \tag{4.17}$$

If we assume $\int_{\sigma} [v] ds \neq 0$ it then follows that

$$\left(\int_{\sigma} [v] ds \right)^2 = I_1^2 + I_2^2 ,$$

and

$$\nu_{0,i} = \pm I_i / \sqrt{I_1^2 + I_2^2} .$$

This gives us a normal vector to the line on which σ lies. To be precise let us choose ν_0 corresponding to the $+$ sign, i.e.,

$$\nu_{0,i} = I_i / \sqrt{I_1^2 + I_2^2} . \tag{4.18}$$

From (4.17) it follows that

$$\int_{\sigma} [v] ds > 0$$

for this choice of ν_0, and therefore

$$\int_{\sigma} [v] ds = \sqrt{I_1^2 + I_2^2} .$$

We now recover the constant c, of the equation $\nu_0 \cdot x = c$ for the line on which σ lies. Let us rotate our coordinate system so that the normal vector ν_0 to σ has coordinates $(0, 1)$. Since we know ν_0 this rotation is known; the line on which σ lies is still $\nu_0 \cdot x = c$, or simply $x_2 = c$. Let

$\phi(x) = \frac{1}{2}(x_2^2 - x_1^2)$ in this new coordinate system. The function $\phi(x)$ is again harmonic. Next define I_ϕ by

$$I_\phi = \int_{\partial\Omega} \left(v\frac{\partial\phi}{\partial\nu} - \phi\frac{\partial v}{\partial\nu} \right) ds \ . \tag{4.19}$$

Application of the divergence theorem gives

$$I_\phi = \int_\sigma [v]\frac{\partial\phi}{\partial\nu}\, ds \ .$$

Since $\nabla\phi = (-x_1, x_2)$ and since $\nu_0 = (0, 1)$, this becomes

$$I_\phi = \int_\sigma [v]x_2\, ds \ ,$$

or

$$I_\phi = c \int_\sigma [v]\, ds \ ,$$

so that

$$c = I_\phi/(\int_\sigma [v]\, ds) = I_\phi/\sqrt{I_1^2 + I_2^2} \ ,$$

again assuming that the denominator does not vanish. This formula in combination with the formulae (4.16) and (4.18) gives a simple way of determining the line on which a single linear, perfectly insulating crack lies.

We cannot generally take for granted that $\int_\sigma [v]\, ds$ is non-zero, given a single fixed boundary current $\frac{\partial v}{\partial\nu}$ on $\partial\Omega$. This is certainly the generic case, but it is not difficult to see that a single applied boundary current can not guarantee that this is true for all σ. This is consistent with the general result that one needs two applied currents and corresponding voltage measurements even to find a single crack. If the applied currents are all two-electrode currents it is very easy to see that a finite number will suffice in order to obtain $I_1^2 + I_2^2 \neq 0$ for any interior crack for at least one of these currents. By a more detailed analysis it is apparently possible to show that any two different two-electrode currents always suffice [2].

We note that the choices for the three auxiliary functions used for reconstructing the line on which the crack lies are harmonic polynomials of degree ≤ 2, quite similar to the weight functions used in our crack reconstruction algorithm. In the limit (as crack length goes to zero) our four weight functions $\omega^{(1)}$ through $\omega^{(4)}$ span the space of harmonic polynomials of degree ≤ 2, modulo constants. This is also the reason why in the case of a disk the limiting information contained in the functionals is equivalent to the first 4 nontrivial Fourier modes, as noted earlier.

Finally let us note that it is very easy to write the quantities I_i and I_ϕ in terms of data corresponding to the harmonic conjugate to v, *i.e.*, in terms of data corresponding to a perfectly conducting σ. Given a smooth function f on $\partial\Omega$ which satisfies $\int_{\partial\Omega} f\, ds = 0$, let $\int^* f$ denote a counterclockwise integral of f along $\partial\Omega$, *i.e.*, a function whose counterclockwise tangential derivative equals f. From the relation (2.3) with $\gamma = 1$ we now immediately obtain

$$I_\phi = \int_{\partial\Omega} \left(v\frac{\partial\phi}{\partial\nu} - \phi\frac{\partial v}{\partial\nu} \right) ds = \int_{\partial\Omega} \left(-\frac{\partial u}{\partial\nu}(\int^* \frac{\partial\phi}{\partial\nu}) - \frac{\partial\phi}{\partial\tau}u \right) ds$$

where $\frac{\partial}{\partial\tau}$ denotes the counterclockwise tangential derivative on $\partial\Omega$. Similar formulae for I_i are obtained by replacing ϕ with ϕ_i. We use these formulae directly instead of making any transformations to our measured boundary data.

158

5 Reconstruction Experiments

We have performed numerous experiments with the equipment described in section 3 followed by reconstructions using the algorithm described in section 2. In this section we discuss the results in three different scenarios, that we find to be representative. In the first scenario two sheet metal strips were inserted in the tank to simulate perfectly conducting cracks. We varied the locations of the two metal strips inside the tank in order to give some idea of how the information contents of the measured data deteriorate as the cracks lie deeper and deeper inside the domain. In the second scenario three sheet metal strips were inserted to demonstrate that the data do indeed permit one to distinguish more than one or two cracks. We have not at the moment experimented with a number of cracks larger than three, partially because of the practical difficulties associated with the accurate insertion of a large number of cracks, but also because we feel it is likely that the currently available accuracy of the data (and the mathematical model) will make such reconstructions extremely crude. In the third scenario we used a single metal strip, but this time bent so that its cross section forms an L-shape. This experiment demonstrates that the collected data have sufficient information contents to distinguish a shape which differs from a line segment, and it also demonstrates the flexibility of the reconstruction algorithm to approximate such a shape. In our opinion the results presented here clearly demonstrate the viability of impedance imaging as a method to reconstruct a moderate number of macroscopic cracks from experimental data.

Let us give a brief description of the figures. Each figure depicts two disks to illustrate one iteration of the reconstruction algorithm. The location of the reconstructed crack(s) are indicated by solid lines. Those in the disk to the left are the locations prior to the particular iteration, and those in the disk on the right are the locations after the iteration is complete. The dashed lines in the disk to the right indicate the locations of the "true" cracks as read off from the mesh at the bottom of the tank. The "optimally sensitive" electrode locations that were used for this iteration are indicated on the boundary of the disk to the left. There is one pair of "optimally sensitive" electrodes corresponding to each reconstructed crack. These electrode pairs are selected according to the strategy briefly described earlier in section 2. n reconstructed cracks will in general correspond to 2n active electrode locations. We do permit some of the active electrode locations to coincide – though we always insist that the n prescribed two- electrode boundary currents be linearly independent (so there must be at least $n + 1$ distinct active electrodes). This strategy is different from that of [7], where we also selected one pair of electrodes corresponding to each crack but fixed one electrode location to be shared by all the electrode pairs; in those reconstructions n cracks corresponded to exactly $n+1$ active electrode locations. We found that in the presence of the "noisy" experimental data the reconstructions were improved by not forcing one common electrode. Although we have not indicated on the figures which electrode pair corresponds to which crack, it is almost evident, given the observation we made earlier concerning a single perfectly conducting crack: the optimally sensitive electrodes are generally located near the intersection of the boundary and the line on which the crack lies. Figure 3a shows the first iteration using experimental data corresponding to two cracks, but in this case the algorithm seeks to reconstruct only a single crack. The initial guess shown in the disk to the left lies on the line generated by the formulae of Andrieux and Ben Abda. We have chosen the initial guess to be of length $0.1 \times$ the radius of the disk, and to be centered on that segment of the line which lies inside the disk. Figure 3b shows the fifth iteration of the algorithm when starting with the initial guess shown in Figure 3a. At this point the algorithm has converged in the sense that the maximum of all four components of G is less than 10^{-10} in absolute value. We now take this crack and divide it into two by omitting a central piece, 1/10 of its size. The resulting two cracks are given as initial guess to our algorithm for the reconstruction of two cracks. After 83 iterations the reconstructed cracks have converged to the two cracks shown in the right disk of Figure 3c. We have run several examples with data originating from two cracks. As one would expect the reconstructed cracks deviate more from the "true" cracks when the "true" cracks are closer to the center of the disk. The above reconstruction is quite representative of cracks located about 1/3 of the way towards the center. Figures 4a and 4b show the iterations at which convergence is reached when reconstructing one and two cracks respectively, given data that correspond to two "true" cracks located nearly halfway towards the center. Figures 5a and 5b show reconstructions obtained from data corresponding to two "true" cracks located approximately 3/4 of the way towards the center. Figure 5a shows the fifth iteration reconstructing only a single crack, at which point this process has converged. We divide the resulting crack into two pieces by cutting out a central piece 1/10 of its size, and feed these two cracks as initial guess to our algorithm reconstructing two cracks. Figure 5b shows the 90th iteration of the two-crack reconstruction process. There is no convergence, nor does any converged state appear to be reached at a later iteration. The collected data is apparently not accurate enough to distinguish these two cracks from a single crack near the center.

We now proceed with a scenario involving three "true" cracks. We start by reconstructing one crack which, in the sense of our functionals, is consistent with the data. The initial guess is generated by means of the Andrieux-Ben Abda formulae, as described earlier. We take the converged crack as shown in Figure 6a (after 6 iterations), divide it into two and feed the resulting cracks as an initial guess to our algorithm for the reconstruction of two cracks. After convergence (in this case after another 33 iterations, as shown in Figure 6b) we divide the largest of the reconstructed cracks into two pieces, and feed the 3 resulting cracks as an initial guess to our algorithm for the reconstruction of three cracks. Figure 6c shows the results of the three-crack reconstruction after 90 iterations. Convergence has not been attained, but notice that the reconstructed cracks none the less give a reasonable prediction of the "true" locations. Also notice that there are only 5 active electrodes on the left circle in Figure 6c. This is therefore an example where two active electrode locations coincide, as discussed earlier.

Finally, in Figures 7a and 7b we show the converged reconstructions using one and two cracks respectively and data which come from insertion of a single strip of metal which has been bent so that its cross section forms an L-shape.

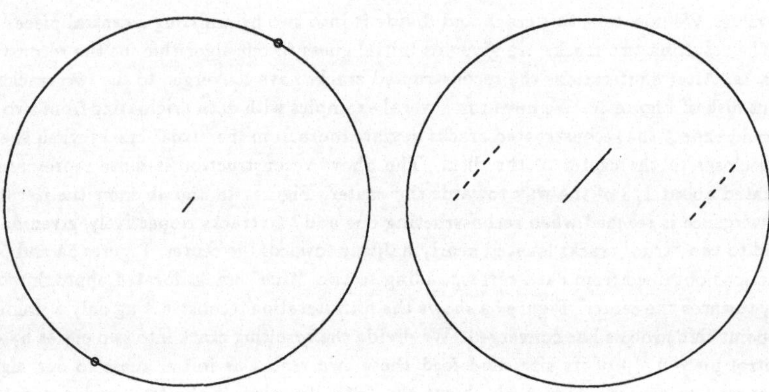

Figure 3a

The first iteration reconstructing one crack. The "true" crack locations are shown in dashed lines in the disk to the right. The currently active electrode locations are indicated on the circle to the left.

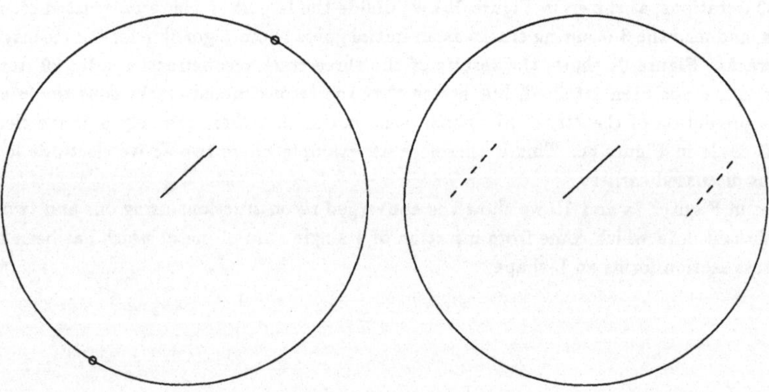

Figure 3b

The 5th iteration reconstructing one crack. The collected data are the same as in Figure 3a. The "true" crack locations are shown in dashed lines in the disk to the right. The currently active electrode locations are indicated on the circle to the left.

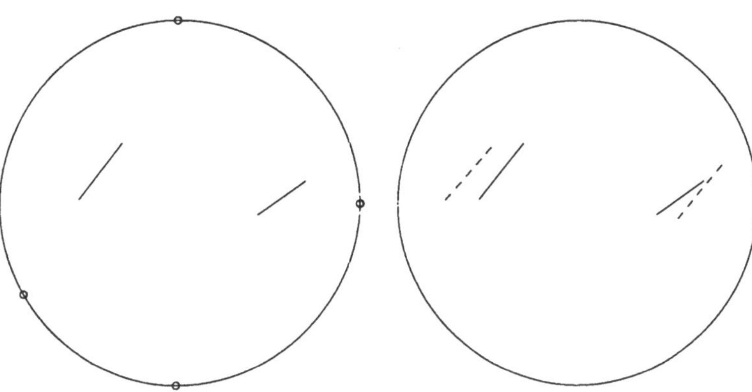

Figure 3c

The 83rd iteration reconstructing two cracks. The collected data are the same as in Figures 3a and 3b. The "true" crack locations are shown in dashed lines in the disk to the right. The currently active electrode locations are indicated on the circle to the left.

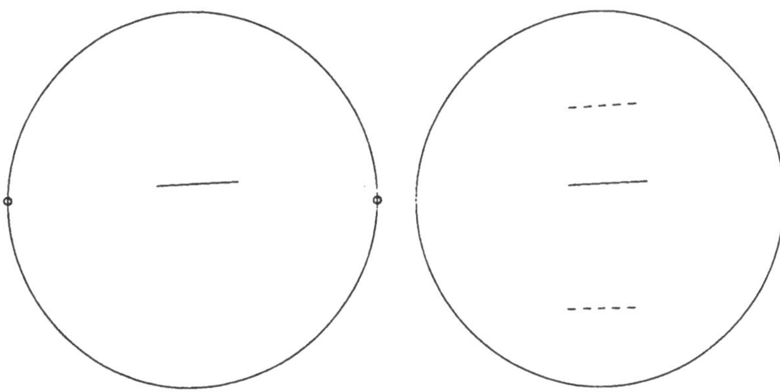

Figure 4a

The 6th iteration reconstructing one crack. The "true" crack locations are shown in dashed lines in the disk to the right. The currently active electrode locations are indicated on the circle to the left.

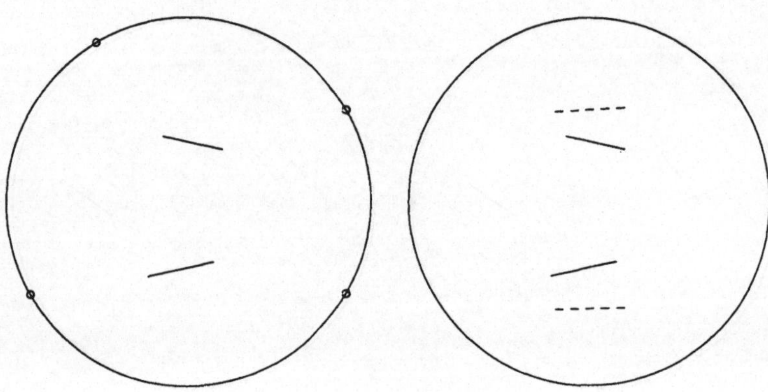

Figure 4b

The 31st iteration reconstructing two cracks. The collected data are the same as in Figure 4a. The "true" crack locations are shown in dashed lines in the disk to the right. The currently active electrode locations are indicated on the circle to the left.

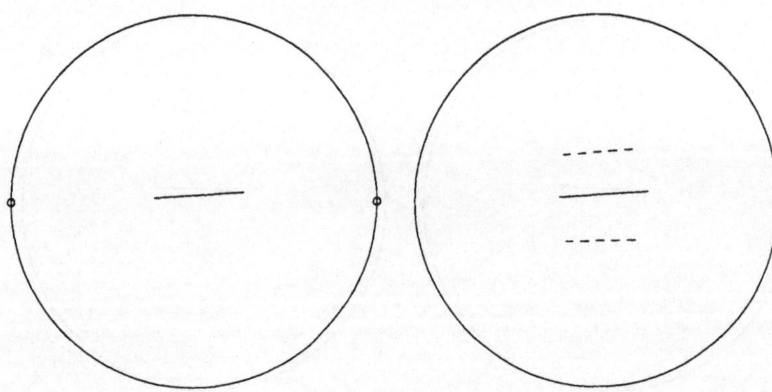

Figure 5a

The 5th iteration reconstructing one crack. The "true" crack locations are shown in dashed lines in the disk to the right. The currently active electrode locations are indicated on the circle to the left.

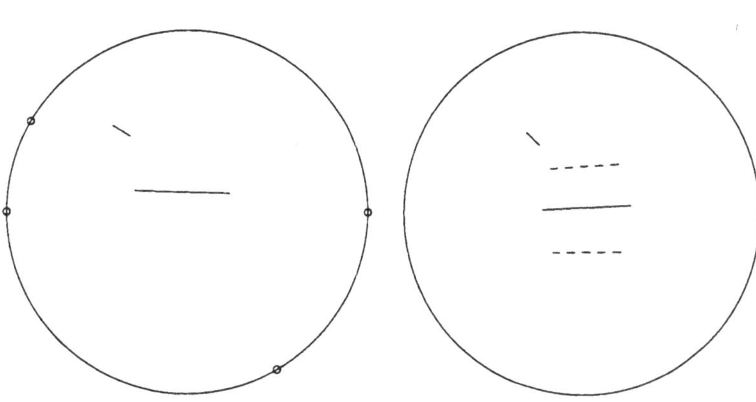

Figure 5b

The 90th iteration reconstructing two cracks. The collected data are the same as in Figure 5a. Convergence has not been reached. The "true" crack locations are shown in dashed lines in the disk to the right. The currently active electrode locations are indicated on the circle to the left.

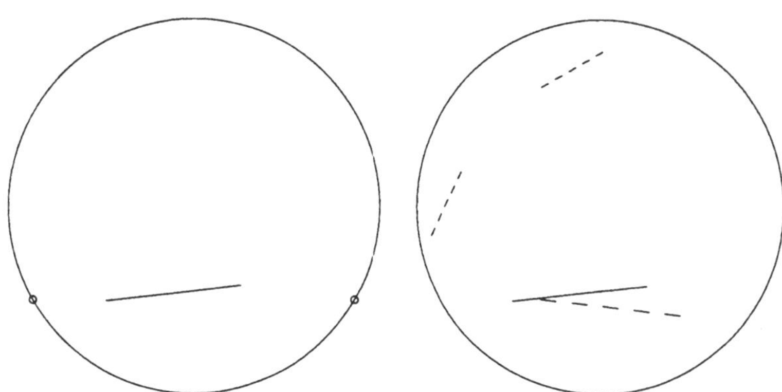

Figure 6a

The 6th iteration reconstructing one crack. The "true" crack locations are shown in dashed lines in the disk to the right. The currently active electrode locations are indicated on the circle to the left.

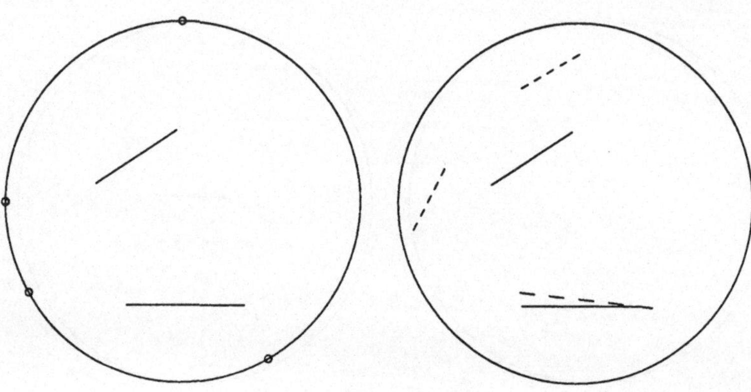

Figure 6b

The 33rd iteration reconstructing two cracks. The collected data are the same as in Figure 6a. The
"true" crack locations are shown in dashed lines in the disk to the right. The currently active
electrode locations are indicated on the circle to the left.

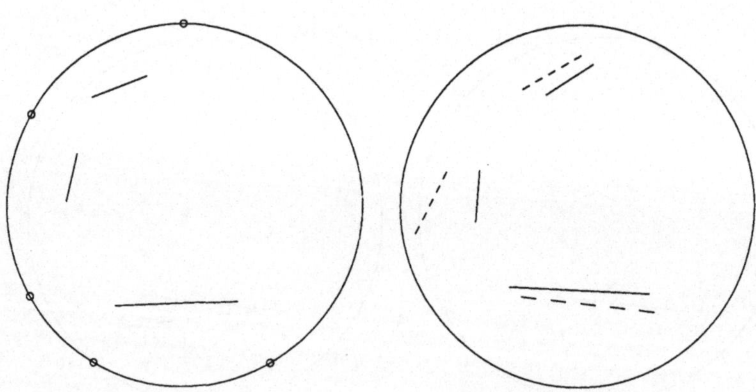

Figure 6c

The 90th iteration reconstructing three cracks. The collected data are the same as in Figures 6a and
6b. Convergence has not been reached. The "true" crack locations are shown in dashed lines in the
disk to the right. The currently active electrode locations are indicated on the circle to the left.

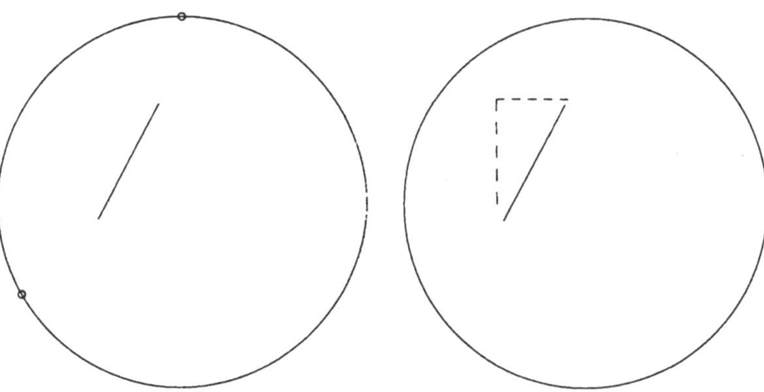

Figure 7a

The 7th iteration reconstructing one crack. The "true" crack location is shown as a dashed curve in the disk to the right. The currently active electrode locations are indicated on the circle to the left.

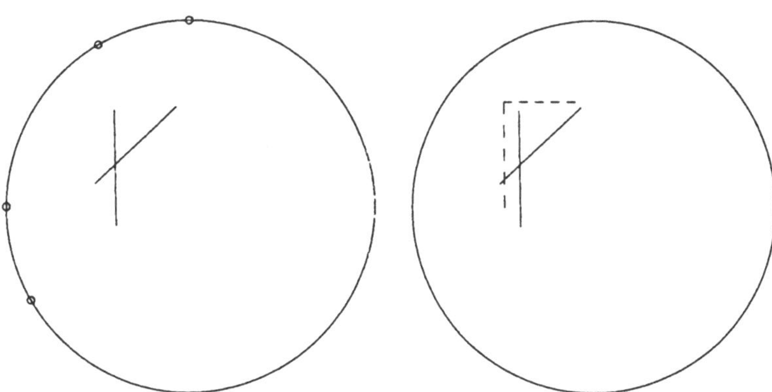

Figure 7b

The 12th iteration reconstructing two cracks. The collected data are the same as in Figure 7a. The "true" crack location is shown as a dashed curve in the disk to the right. The currently active electrode locations are indicated on the circle to the left.

Acknowledgments

This research was partially supported by NSF grant DMS-9202042 and by AFOSR contract 89NM605. This work was performed while the first author was in residence at the Institute for Computer Applications in Science and Engineering (ICASE), NASA Langley Research Center, Hampton, VA 23681, which is operated under National Aeronautics and Space Administration contracts NAS1-18605 and NAS1-19480. This work was performed while the third author was visiting the Université Joseph Fourier during the academic year 92/93. He would like to thank his colleagues at "Laboratoire de Modelisation et Calcul" for the hospitality extended to him and in particular Jacques Blum for his efforts towards making this visit possible.

References

[1] Alessandrini, G., Stable determination of a crack from boundary measurements. Proc. Roy. Soc. Edinburgh, Section A **123** (1993), pp. 497–516.

[2] Alessandrini, G., Personal communication.

[3] Alessandrini, G., Beretta, E. and Vessella, S., Determining linear cracks by boundary measurements – Lipschitz stability. Preprint, 1993.

[4] Andrieux, S. and Ben Abda, A., Identification de fissures planes par une donnée de bord unique: un procédé direct de localisation et d'identification. C. R. Acad. Sci. Paris, Serié I, **315** (1992), pp. 1323–1328.

[5] Barber, D. and Brown, B., Recent developments in applied potential tomography - APT, in Information Processing in Medical Imaging, Bacharach, S. ed., Nijhoff, Amsterdam, 1986, pp. 106–121.

[6] Bryan, K. and Vogelius, M., A uniqueness result concerning the identification of a collection of cracks from finitely many electrostatic boundary measurements. SIAM J. Math. Anal., **23** (1992), pp. 950–958.

[7] Bryan, K. and Vogelius, M., A computational algorithm to determine crack locations from electrostatic boundary measurements. The case of multiple cracks. To appear in Int. J. Engng. Sci.

[8] Cheney, M., Isaacson, D., Newell, J., Simske, S. and Goble, J., NOSER: An algorithm for solving the inverse conductivity problem, Int. J. Imaging Systems and Tech., **22** (1990), pp. 66–75.

[9] Diaz Valenzuela, A. Unicità et stabilità per il problema inverso del crack perfettamente isolante. Thesis, University of Trieste, 1993.

[10] Eggleston, M.R., Schwabe, R.J., Isaacson, D., Goble, J.C. and Coffin, L.F., Three-dimensional defect imaging with electric current computed tomography. GE Technical Report 91CRD039, Schenectady, NY.

[11] Friedman, A. and Vogelius, M., Determining cracks by boundary measurements. Indiana Univ. Math. J., **38** (1989), pp. 527–556.

[12] Gisser, D.G., Isaacson, D. and Newell, J.C., Electric current computed tomography and eigenvalues. SIAM J. Appl. Math., **50** (1990), pp. 1623–1634.

[13] Liepa, V., Santosa, F. and Vogelius, M. Crack determination from boundary measurements – Reconstruction using experimental data. J. Nondestructive Evaluation, **12** (1993), pp. 163–174.

[14] Moré, J., The Levenberg-Marquardt algorithm: implementation and theory. *Numerical Analysis* (Edited by Watson, G.A.), pp. 105–116. *Lecture Notes in Math.* 630. Springer Verlag, 1977.

[15] Nishimura, N., Regularized integral equations for crack shape determination problems. *Inverse Problems in Engineering Sciences* (Edited by Yamaguti, M. et al), pp. 59–65. Springer Verlag, 1991.

[16] Nishimura, N. and Kobayashi, S., A boundary integral equation method for an inverse problem related to crack detection. Int. J. Num. Meth. Eng., **32** (1991), pp. 1371–1387.

[17] Santosa, F. and Vogelius, M., A computational algorithm to determine cracks from electrostatic boundary measurements. Int. J. Eng. Sci. **29** (1991), pp. 917–937.

[18] Yorkey, T., Webster, J. and Tompkins, W., Comparing reconstruction algorithms for electrical impedance tomography. IEEE Trans. Biomedical Eng., **BME-34** (1987), pp. 843–852.

Department of Mathematics
Rose-Hulman Institute of Technology
Terre Haute, IN 47803

Radiation Laboratory, Dept EECS
University of Michigan
Ann Arbor, MI 48109

Department of Mathematics
Rutgers University
New Brunswick, N.J. 08903

Monitoring of Transient Temperature Distribution in Piping

K. BABA and M. OCHI

Abstract

A numerical method for monitoring temperature distribution in which boundary flux and initial state are unknown is presented. Regularizations based on Tikhonov's and Beck's method are employed. And then, regularization parameters are evaluated by *L-curve*. The method is applied to an actual piping problem in a steam power plant and compared with measured data, and it is also applied to a two-dimensional thermal shock problem.

1 Introduction

In this paper a monitoring problem in a piping system is described. In steam and nuclear power plants themal changes during starting and stopping give rise to thermal stress in the piping. Emergency cooling water used in case a nuclear power plant accident causes therml shock. To detect fatigue damage in plant piping, it is necessary to monitor the temperature distributions in the interior. Determing temperatue ditributions from data given at certain parts is called the inverse heat conduction problem (IHCP). IHPC is studied in many applications, for examples see : [4], [7],[16]. Many numerical and mathematical studies were also done , for examples see : [2],[3],[5],[8], [9],[10],[11],[13],[15]. These studies are being conducted to determine only initial conditions or boundary conditions assuming other conditions for each. The purpose of the inverse problem that is shown in this paper is to monitor temperature under more general conditions. A numerical method that identifies initial distribution and boundary flux by measuring the surface is presented. The solution to this problem is not unique. An example of application to an actual plant is shown and compared with measured data. The results of two dimensional thermal shock problem are also shown.

2 Problem

The purpose of this paper is to determine the transient temperature distributions throughout a pipe by suface temperature measurement. This problem is written as follows,

Problem 2.1 *Find u, q, f such that*

$$\frac{\partial u}{\partial t} = \nabla k \cdot \nabla u \qquad in\ \Omega$$

$$u(t,x)|_{\Gamma_o} = u_o(t,x)$$

$$\frac{\partial u}{\partial n}|_{\Gamma_o} = 0$$

$$\frac{\partial u}{\partial n}|_{\Gamma_q} = q(t,x)$$

$$u(0,x) = f(x) \qquad in\ \Omega$$

where $k > 0$ is a physical parameter and u_o is a given function.

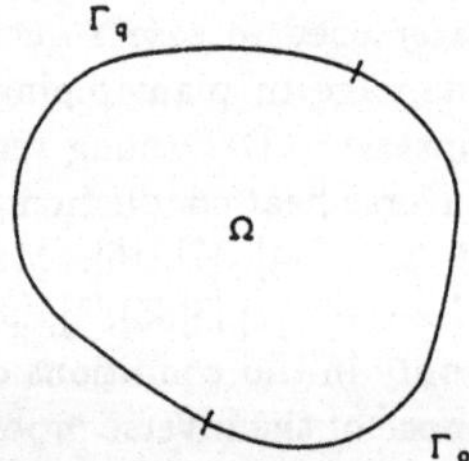

Figure 1: Domain and notations

Remark 2.1 *The solution of Problem 2.1 is not unique.*

For simplicity let us assume that k is constant, using the Green function G the solution u is expressed as

$$u(t,x) = \int_\Omega G(t,x;0,\xi)f(\xi)d\xi - k\int_0^t \int_{\Gamma_o} G(t,x;\tau,\xi)q(\tau,\xi)d\xi d\tau$$

We represent the solution u as the sum of u_I and u_B which is given by,

$$u_I(t, x) = \int_\Omega G(t, x; 0, \xi) f(\xi) d\xi$$

$$u_B(t, x) = -k \int_0^t \int_{\Gamma_e} G(t, x; \tau, \xi) q(\tau, \xi) d\xi d\tau$$

Each of them has initial and boundary conditions as follows,

$$u_I(0, x) = f(x)$$

$$\frac{\partial u_I}{\partial n}\Big|_{\Gamma_o \cup \Gamma_e} = 0$$

$$u_B(0, x) = 0$$

$$\frac{\partial u_B}{\partial n}\Big|_{\Gamma_e} = q(t, x)$$

$$\frac{\partial u_B}{\partial n}\Big|_{\Gamma_o} = 0$$

Remark 2.1 can be shown easily as follows. For any boundary flux $q' \neq 0$ on Γ_o we can obtain u'_B such that,

$$\frac{\partial u'_B}{\partial t} = \nabla k \cdot \nabla u'_B \qquad \text{in } \Omega$$

$$\frac{\partial u'_B}{\partial n}\Big|_{\Gamma_o} = 0$$

$$\frac{\partial u'_B}{\partial n}\Big|_{\Gamma_e} = q'$$

$$u'_B(0, x) = 0 \qquad \text{in } \Omega$$

Following the direct problem above, we consider the following inverse problem.

Problem 2.2 *Find u'_I and f' such that,*

$$\frac{\partial u'_I}{\partial t} = \nabla k \cdot \nabla u'_I \qquad \text{in } \Omega$$

$$u'_I(0, x) = f'(x) \qquad \text{in } \Omega$$

$$\frac{\partial u'_I}{\partial n}\Big|_{\Gamma_o \cup \Gamma_e} = 0$$

$$u'_I\big|_{\Gamma_o} = -u'_B\big|_{\Gamma_o}$$

The observability of Problem 2.2 was studied by Sakawa [13]. The function $u' = u'_I + u'_B$ is the solution of Problem 2.1 with the observed data $u_o = 0$. Then Problem 2.1 does not have a unique solution.

Inversely, for any initial distribution $f'' \neq 0$ in Ω we consider an initial value problem,

$$\frac{\partial u''_I}{\partial t} = \nabla k \cdot \nabla u''_I \quad \text{in } \Omega$$

$$\frac{\partial u''_I}{n}\big|_\Gamma = 0$$

$$u''_I(0, x) = f'' \quad \text{in } \Omega$$

Following this direct problem, we consider the following inverse problem.

Problem 2.3 *Find $u''_B and q''$ such that,*

$$\frac{\partial u''_B}{\partial t} = \nabla k \cdot \nabla u''_B \quad in \ \Omega$$

$$u''_B(0, x) = 0 \quad in \ \Omega$$

$$\frac{\partial u''_B}{n}\big|_{\Gamma_o} = 0$$

$$\frac{\partial u''_B}{n}\big|_{\Gamma_q} = q''$$

$$u''_B\big|_{\Gamma_o} = -u''_I\big|_{\Gamma_o}$$

Let $V = V_f \times V_q$ be the kernel of Problem 2.1. The above discussion means that under no constraint, can any of V_f or V_q be a solution to Problem 2.1. And when conditions for the uniqueness of Problem 2.2 and 2.3 are assumed V_f and V_q is one to one.

3   Numerical method

We apply the least square method, for observed data u_o

$$\Pi_o = \int_0^t \int_{\Gamma_o} (u - u_o)^2 ds dt$$

The solution u can be represented by a sum of u_I and u_B. For the sake of generality the fundamental solutions for each are obtained numerically by FEM.

3.1 Fundamental solutions for u_I

Let us consider an eigenvalue problem such that,

$$\nabla k \cdot \nabla \phi_i = \lambda_i \phi_i \qquad \text{in } \Omega$$

$$\frac{\partial \phi_i}{\partial n}\Big|_{\Gamma_o \cup \Gamma_\ell} = 0,$$

$$i = 1, 2, \cdots, N_I$$

By using these eigenvalues and eigenfunctions, the fundmental solutions are written as follows,

$$\phi_i e^{-\lambda_i t}, \quad i = 1, 2, \cdots, N_I$$

So u_I is now expressed as,

$$u_I = \sum_{i=1}^{N_I} a_i \phi_i(x) e^{-\lambda_i t}$$

3.2 Fundamental solutions for u_B

The fundamental solutions for u_B are obtaind by direct solution.

$$\frac{\partial \psi_i}{\partial t} = \nabla k \cdot \nabla \psi_i \qquad \text{in } \Omega$$

$$\psi_i(0, x) = 0 \qquad \text{in } \Omega$$

$$\frac{\partial \psi}{\partial n}\Big|_{\Gamma_o} = o$$

$$\frac{\partial \psi}{\partial n}\Big|_{\Gamma_\ell} = q_i(s)$$

$$i = 1, 2, \cdots, M_B$$

where the flux $q_i, i = 1, 2, \cdots, M_B$, are linear independent functions. We denote ψ_i^j as,

$$\psi_i^j(t, x) = \begin{cases} \psi_i(t - t_j, x), t \geq t_j \\ 0, t < t_j \end{cases} \quad , j = 1, 2, \cdots, N_B$$

Using these functions we can write u_B as follows,

$$u_B(t, x) = \sum_{i=1}^{M_B} \sum_{j=1}^{N_B} b_i^j \psi_i^j(t, x)$$

3.3 Regularization

We apply the Tikhonov regularization method. Adding the regularization terms to Π_0 we obtain a functional,

$$\Pi = \Pi_0 + \alpha_1 \int_0^t \int_{\Gamma_q} \left(\frac{\partial q}{\partial s}\right)^2 ds dt + \alpha_2 \int_0^t \int_{\Gamma_q} \left(\frac{\partial q}{\partial t}\right)^2 ds dt$$

$$+ \alpha_3 \int_\Omega \nabla f \cdot \nabla f dx + \alpha_4 \int_\Omega (\Delta f)^2 dx$$

where $\Delta f = u(\Delta t, x) - f(x)$ and $\alpha_i, i = 1, 2, 3$, mean regularization parameters.

The regularization terms on Γ_q are similar to [14]. The last two terms on the right hand side fill the essential role to determine the initial distribution and the boundary flux at the first time step. We seek a solution (a_i, b_j^k) which minimizes the functional Π through an iterative procedure with initial distribution, a_i, and boundary flux, b_j^k, of several time steps.

3.4 Noise of observed data and regularization parameters

For simplicity, we write the problem in the following form.

$$Au = b$$

where b means the observed data. Then the regularized functional is now expressed as,

$$F_\alpha^\delta = \parallel Au - b^\delta \parallel^2 + \alpha \parallel u \parallel^2$$

where b^δ denotes the observed data that include noise Δb.

$$b^\delta = b + \Delta b$$

Let δ be a constant that satisfies the following inequality.

$$\parallel \Delta b \parallel \leq \delta \leq \parallel b^\delta \parallel$$

We denote u_0^δ the least square least norm solution with the observed data b^δ.

$$\parallel u_0^\delta \parallel = \min_{u \in U} \parallel u \parallel$$

$$U \underset{def}{=} \left\{ u : \parallel Au - b^\delta \parallel = \min_{x \in V} \parallel Ax - b^\delta \parallel \right\}$$

where V is an appropriate function space.

The discrepancy principle by Morozov claims, for u_α^δ that is the least square solution of F_α^δ,

$$\| u_0 - u_\alpha^\delta \| \leq E_\alpha^\delta$$

$$E_\alpha^\delta \underset{def}{=} \| u_\alpha^\delta \|^2 - \frac{2}{\alpha}(r_\alpha^\delta, b^\delta) + \frac{2\delta}{\alpha} \| r_\alpha^\delta \| + \| u_0 \|^2$$

where r_α^δ is the residual which is defined as,

$$r_\alpha^\delta \underset{def}{=} Au_0^\delta - Au_\alpha^\delta$$

Furthermore, E_α^δ takes the minimum when $\| r_\alpha^\delta \| = \delta$.

3.5   Evaluation of the regularization parameters

In practice the noise of the data Δb cannot be known. It is important how we evaluate the value of the regularization parameter α.

Hansen proposed *L-curve* which is defined by the axes $\| u_\alpha^\delta \|$ and $\| r_\alpha^\delta \|$. The *L-curve* is a monotone function and it tends to $(0, \| u_0^\delta \|)$ as $\alpha \to 0$ and $(\| b_0^\delta \|, 0)$ as $\alpha \to \infty$ (Figure 2).

Hosada and Kitagawa [6] proposed to use α which maximizes the curvature.

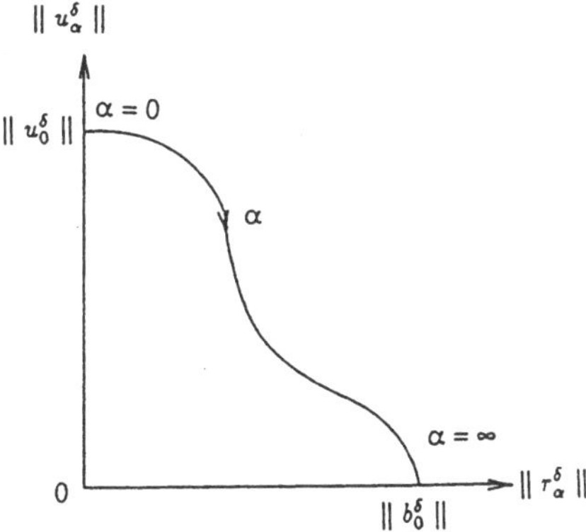

Figure 2: *L-curve*

3.6 Sequential procedure

The number of the unknowns, b_i^j, increases in proportion to the number of time steps. Then after getting the initial distribution and boundary fluxes of several time steps as written in the previous sections, we apply Beck's method [1], which is a sequential regularization method to determine boundary conditions. For known u^n and q^n , where n means a time step ,let $u_1^{n+i}, i = 1, 2, \cdots, m$, be the solutions to the direct problem with the initial flux q^n. In the same way let $u_2^{n+i}, i = 1, 2, \cdots, m$ be the solution with unit flux and initial condition free. When the flux of n+1 step is,

$$q^{n+1} = q^n + \Delta q$$

the solution with q^{n+1} is written as,

$$u^{n+i} = u_1^{n+i} + \Delta q\, u_2^{n+i}, i = 1, 2, \cdots, m$$

For the observed data $u_o^{n+i}, i = 1, 2, \cdots, m$, the least square functional,

$$\Pi = \sum_{i=1}^{m}(u_o^{n+i} - u^{n+i})^2$$

is minimized by,

$$\Delta q = \frac{\left\{ \sum_{i=1}^{m} u_2^{n+i}(u_o^{n+i} - u_1^{n+i}) \right\}}{\left\{ \sum_{i=1}^{m} (u_2^{n+i})^2 \right\}}$$

This method leads to a regularization of the problem by $\| \frac{\partial q}{\partial t} \|$. We apply this method and a spatial regularization to multi-dimensional problems.

4 Numerical example

We show an example of application in a piping of a steam power plant. We applied it to a T-joint of the third pendant super heater (Figure 3). Temperatures are measured at three points in the direction of thickness. We used data from a warm start operation. The outside surface condition is nearly flux free. The material is SA335-P91 and its constants are shown in Table 1. We solved this as a one dimensional problem. The measured temperatures are shown in Figure 4. We used the temperatures of TmST3.17 as the input data. The number of fundamental solutions for the initial condition is five and that for the boundary flux is one because of the one dimensional problem. The number of time steps for Beck's method is six. Under these conditions we applied the method discussed in the previous sections. The regularization parameters are $\alpha_2 = 10^{-4}$, $\alpha_3 = 10^{-6}$ and $\alpha_4 = 1$. Because this problem is one dimensional, α_1 has no meaning. The time difference is two minutes.

Figure 5 is the result that is applied to the interval from 0 to 30 minutes. And the result that was applied to the interval from 75 to 100 minutes is shown in Figure 6. The marked points are measured temperatures shown in Figure 4. The solid lines in the temperature histories are the values on the inside boundary. These results are sufficient for practical use.

We show a two dimensional thermal shock water flow problem in a heated pipe (Figure 7). We assume outside boundary condition is flux free. We compare the results with that by direct simulation. Outside temperatures are input as measured data. The material constants are shown in Table 2. The regularization parameters are $\alpha_1 = 1, \alpha_2 = 10^{-7}, \alpha_3 = 10^{-6}$ and $\alpha_4 = 1$. And the number of time steps for Beck's method is six. The time difference is 0.5 second. We compared the results of three cases of measured points as shown in Figure 8. The temperature histories for the boundaries by 17 measured points are shown in Figure 9. The influence of the number of measured points is shown in Figures 10 and 11. The result using four points is less accurate. In this problem we need a sufficient number of measured points.

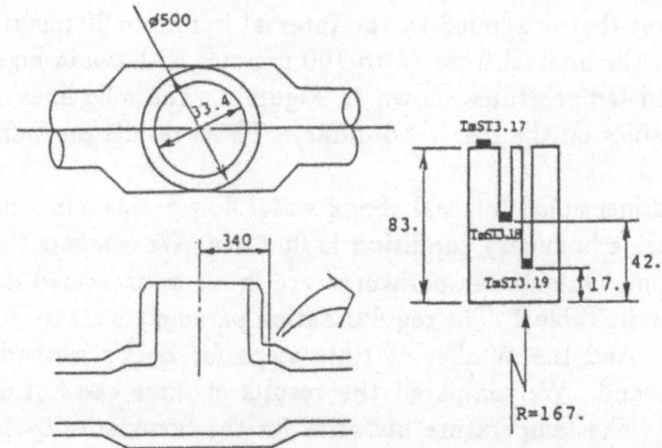

Figure 3: T-joint and measured points

Table 1: Material constants of T-joint

heat conduction	25.0	$kcal/m \cdot h \cdot {}^{\circ}c$
density	7850.0	kg/m^3
specific heat	0.122	$kcal/kg \cdot {}^{\circ}c$

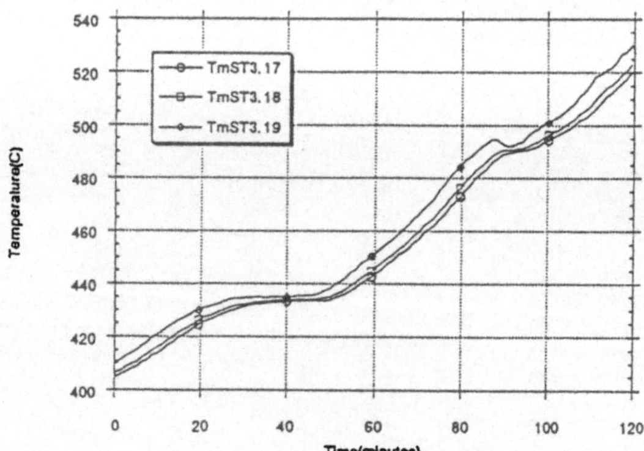

Figure 4: Measured temperatures

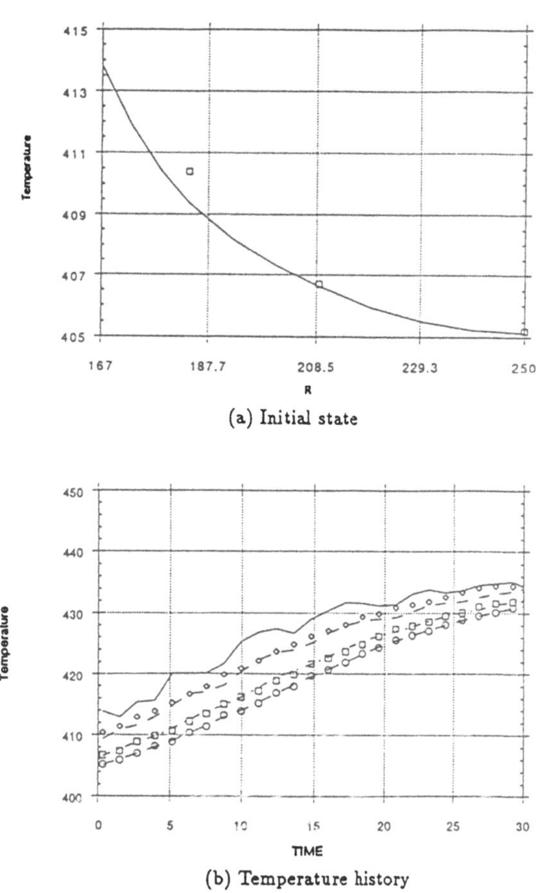

(a) Initial state

(b) Temperature history

Figure 5: Temperature history and distribution started at 0 minute

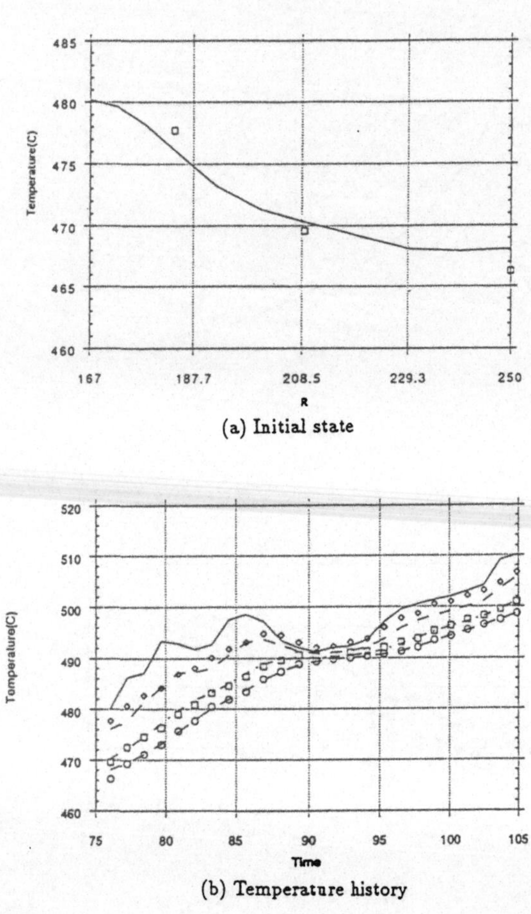

(a) Initial state

(b) Temperature history

Figure 6: Temperature history and distribution started at 75 minutes

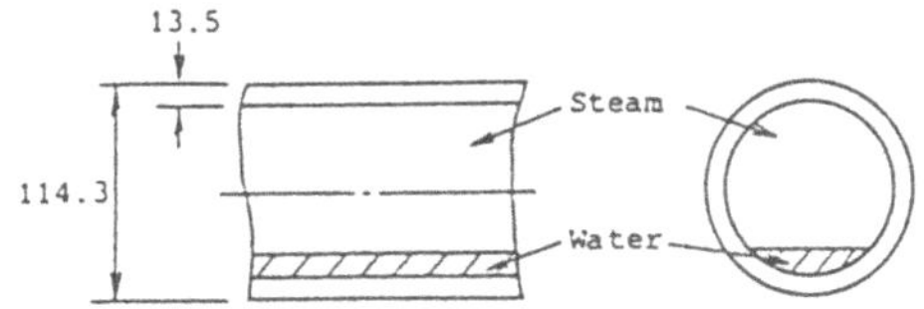

Figure 7: Thermal shock model

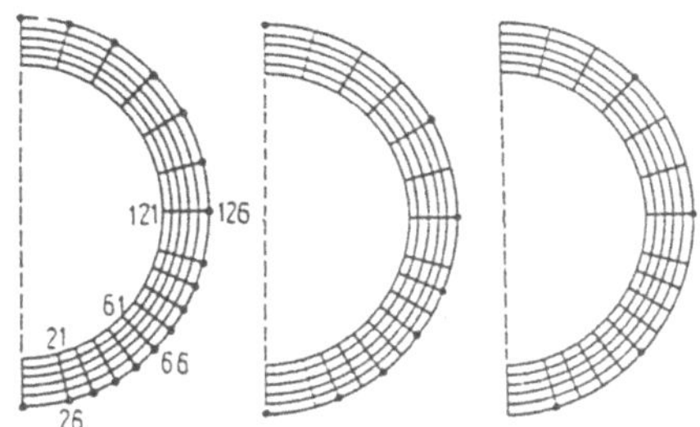

(a) 17 points   (b) 8 points   (c) 4 points

Figure 8: Mesh and measured points

Table 2: Material constants of thermal shock model

heat conduction	16.3	$kcal/m \cdot {}^{\circ}c$
density	8030.0	kg/m^3
specific heat	0.18	$kcal/kg \cdot {}^{\circ}c$

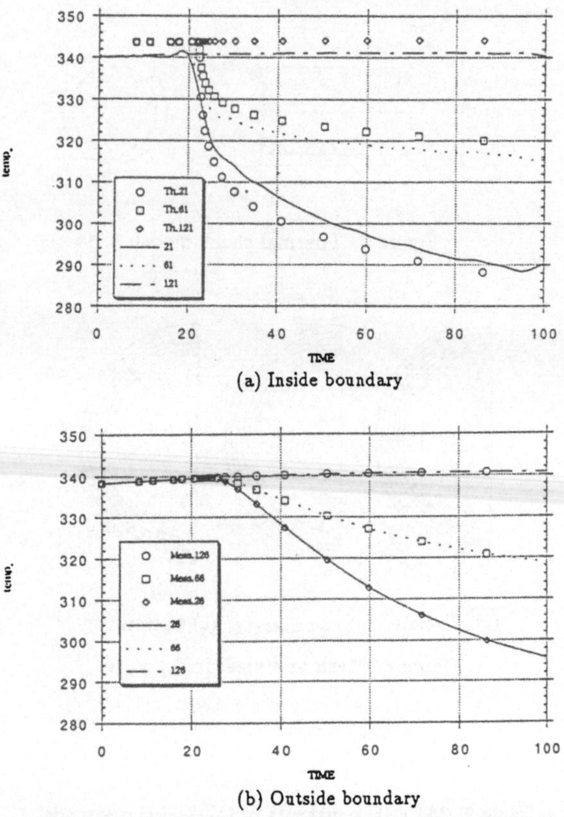

(a) Inside boundary

(b) Outside boundary

Figure 9: Temperature history of thermal shock model

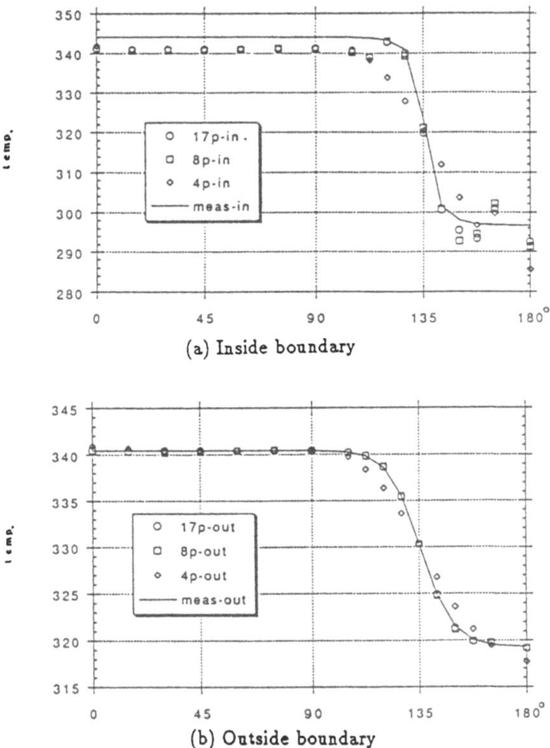

(a) Inside boundary

(b) Outside boundary

Figure 10: Temperature distribution at 50.5 sec of thermal shock model

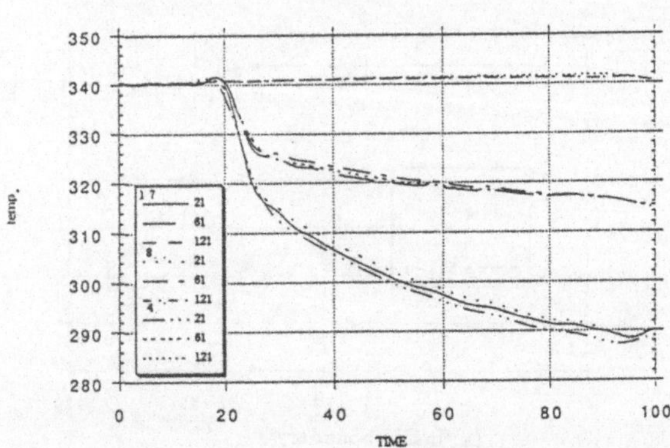

Figure 11: Difference by number of measured points of thermal shock model

References

[1] J.V.Beck,B.Litkouhi,C.R.St.Clair. *Efficient Sequential Solution of the Nonlinear Inverse Heat Conduction Problem*, Numer. Heat Transfer Vol.5 pp.275-286,1982.

[2] H.R.Busby,D.M.Trujillo. *Numerical Solution to a Two-Dimensional Inverse Heat Conduction Problem*, Int. J. Numer. Meth. Engng Vol.21 pp.349-359,1985.

[3] A.S.Carasso. Infinitely Divisible Pulses,Continuous Deconvolution,and the Characterization of Linear Time Invariant Systems, SIAM J. Appl. Math. Vol.47 pp892-927,1987.

[4] H.W.Engl,T.Langthaler. *Numerical Solution of an Inverse Problem Connected with Continuous Casting of Steel*, ZOR-Z. Oper. Res.,Vol.29 B185-B199,1985.

[5] D.Gilliam,Z.Li,C.Martin. *Discrete Observability of the Heat Equation on Bounded Domains*, Int. J. Control Vol.48 pp.755-780,1988.

[6] Y.Hosoda,T.Kitagawa. *Optimal Regularization for Ill-Posed Problem by Means of L-curve*, Transaction of the Japan SIAM Vol.2 pp.55-67,1992 (in Japanese).

[7] Th.Kaiser,F.Troltzsch. *An Inverse Problem Arising in steel cooling Process*, Wiss.Z.Tech.Univ.Karl-Marx-Stadt. Vol.29 pp212-218,1987.

[8] T.Kobayashi. *Initial State Determination for Distributed Parameter Systems*, SIAM J. Control and Optimization Vol.14 pp.934-944,1976.

[9] P.Manselli,K.Miller. *Calculation of the Surface Temperature and Heat Flux on One Side of a Wall from Measurements on Opposite Side*, Ann. Math. Pure Appl. Vol.123 pp161-183,1980.

[10] P.Monk. *Error Estimates for a Numerical Method for an Ill-Posed Cauchy Problem for the Heat Equation*, SIAM J. Numer. Anal. Vol.23 pp1155-1172,1980.

[11] D.A.Murio. *The Mollification Method and the Numerical Solution of the Inverse Heat Conduction Problem by Finite Differences*, Comput. Math. Appl. Vol.17 pp1385-1396,1989.

[12] H.J.Reinhardt. *A Numerical Method for the Solution of Two-dimensional Inverse Heat Conduction Problems*, Int. J. Numer. Meth. Engng Vol.32 pp363-383,1991.

[13] Y.Sakawa. *Observability and Related Problems for Partial Differential Equations of Parabolic Type*, SIAM J. Control and Optimization Vol.13 pp.15-27,1975.

[14] D.S.Schnur,N.Zabaras. *Finite Element Solution of Two-Dimensional Inverse Elastic Problems Using Spatial Smoothing*, Int. J. Numerical Methods in Engineering Vol.30 pp.57-75,1990.

[15] D.F.Tandy,D.M.Trujillo,H.R.Busby. *Solution of Inverse Heat Conduction Problems Using an Eigenvalue Reduction Technique*, Numer. Heat Transfer Vol.10 pp.597-617,1986.

[16] G.J.Weir. *Surface Mounted Heat Flux Sensors*, J. Austral. Math. Soc. Ser.B,Vol.27 pp281-294,1986.

Electronics Division, Mitsubishi Heavy Industries,Ltd.
Wadasaki-chyo Hyogo-ku Kobe 652, JAPAN

Computational Differentiation and Multidisciplinary Design[*]

Christian Bischof[†] and Andreas Griewank[‡]

Abstract

Multidisciplinary Design Optimization (MDO) by means of formal sensitivity analysis requires that each single-discipline analysis code supply not only the output functions for the (usually constrained) optimization process and other discipline analysis inputs, but also the derivatives of all of these output functions with respect to its input variables. Computational differentiation techniques and automatic aifferentiation tools enable MDO by providing accurate and efficient derivatives of computer programs with little human effort. We discuss the principles behind automatic differentiation and give a brief overview of automatic differentiation tools and how they can be employed judiciously, for example, for sparse Jacobians and to exploit parallelism. We show how, and under what circumstances, automatic differentiation applied to iterative solvers delivers the mathematically desired derivatives. We then show how derivatives that can now be feasibly obtained by computational differentiation techniques can lead to improved solution schemes for nonlinear coupled systems and multidisciplinary design optimization.

1   Introduction

Computational differentiation (CD) provides a new foundation for sensitivity analysis and subsequent design optimization of complex systems by reliably computing derivatives of large computer codes. The goal is to free the computational scientist from worrying about the accurate and efficient computation of derivatives, even for complicated "functions," thereby enabling him to concentrate on the more important issues of system modeling and algorithm design.

Almost as soon as the first programmable systems became available, scientists observed that the mechanical application of the chain rule to obtain

---

[*]This work was supported by the Office of Scientific Computing, U.S. Department of Energy, under Contract W-31-109-Eng-38, and by the National Aerospace Agency under Purchase Order L25935D.

[†]Mathematics and Computer Science Division, Argonne National Laboratory, 9700 S. Cass Ave., Argonne, IL 60439, U.S.A., `bischof@mcs.anl.gov`.

[‡]Institute of Scientific Computing, Technical University Dresden, Mommsenstr. 13, 01062 Dresden, Germany, `griewank@math.tu-dresden.de`.

derivatives can be *automatically* and reliably performed by computers. The basic *forward*, or *bottom-up*, mode of automatic differentiation (AD) was apparently first proposed by R. E. Wengert [41]. The mathematically more interesting *reverse*, or *top-down*, mode was first published by G. M. Ostrowskii [36]. This observation gives rise to *automatic differentiation* (AD), which can compute derivatives of a function defined by a computer code in a black-box fashion, without any knowledge of the application beyond what are considered "dependent" and "independent" variables with respect to differentiation.

While these original results have been repeatedly rediscovered and extended, the full potential of automatic differentiation as a general-purpose computational tool has not yet been realized. Historically, this is due both to a lack of general-purpose tools and a lack of scientific communication and cooperation. It was not until 1991 that the first meeting focusing on Computational Differentiation was held [24]. Since then, much has changed as automatic differentiation tools have acquired a level of maturity that enables them to deliver the promises of the theory.

As researchers have gained experience with automatic differentiation tools, it also has become apparent that the process of generating suitable derivative codes is not necessarily automatic. For example, the user of an AD tool can capitalize, in a high-level fashion, on the structure of his program to decrease the computational cost of computing derivatives, leading to an "automated," or "computer-supported" approach to generating derivative code.

Moreover, there is the issue of AD versus "do-what-I-mean," where one has to relate knowledge about the algorithms underlying the program in question with automatic differentiation to ensure that the derivatives computed are the ones that are mathematically desired. Computational differentiation is the term we chose to denote all these closely related efforts.

This paper is structured as follows. In the next section we review automatic differentiation and comment on the mathematical and computer science challenges that remain. In Section 3 we give examples that show how judicious use of AD tools can have a significant effect on the speed with which one computes derivatives. We then explore, in Section 4, whether, and under what circumstances, AD techniques, when applied to iterative solvers, deliver the desired derivatives. Section 5 illustrates how computational differentiation techniques can be employed to arrive at faster and more robust procedures for multidisciplinary design analysis and optimization. Lastly, we give a brief outlook on the field.

2 Automatic Differentiation

Because accurate and efficient derivative values are so crucial for the speed and robustness of numerical methods, many developers have expended great efforts to derive, by hand, code for the derivatives of particular modeling functions. Obviously there would be little point in discussing the importance of AD if there were other approaches for obtaining derivatives accurately and cheaply. We have already mentioned that divided differences are of uncertain quality in terms of accuracy and are also quite costly if there are many independent variables. Fully symbolic differentiation, as provided, for example, by the "diff" operator in MAPLE, is generally not a realistic alternative, because with the possible exception of some right-hand sides in ordinary differential equations, the problem functions in the applications mentioned above are evaluated by computational procedures of some length. Consequently, the direct representation and differentiation of the dependent variables in terms of the independent variables are generally very resource demanding or impossible, while they pose no difficulties for automatic differentiation (see, for example, [9, 28]).

In this section, we give an overview of the classical approaches to automatic differentiation, as well as some of the more recent research questions relating to the computational complexity for computing derivatives. We also give a brief overview automatic differentiation tools, in particular the ADIFOR Fortran77 tool.

2.1 The Forward and Reverse Mode of Automatic Differentiation

In contrast to fully symbolic packages, automatic differentiation applies the chain rule to numbers rather than algebraic expressions. These real numbers are, of course, rounded after each application of the chain rule, so that the complexity of representing and manipulating each partial derivative does not grow at all. It has been shown that automatic differentiation is backward stable in the sense that the derivative values obtained correspond to the exact results for slightly perturbed independent variables and intermediates [26].

As an example let us consider the following simple program with four independent variables x_1, x_2, x_3, x_4 and three dependent variables y_1, y_2, y_3, where the c_i and d_i are distinct constants.

To calculate the Jacobian $\partial y/\partial x$ one could associate with each scalar variable a gradient of derivatives with respect to the three independent variables. Starting from the Cartesian basis vectors $\nabla x_i = e_i \in \mathbf{R}^3$, one can calculate derivatives according to the usual interpretation of the chain rule, as shown in

$$
\begin{aligned}
v_1 &= x_2 * x_3 \\
v_2 &= v_1 x_1 && // && = x_1 x_2 x_3 \\
v_3 &= v_1 x_4 && // && = x_2 x_3 x_4 \\
y_1 &= c_1 v_2 + d_1 v_3 && // && = c_1 x_1 x_2 x_3 + d_1 x_2 x_3 x_4 \\
y_2 &= c_2 v_2 + d_2 v_3 && // && = c_2 x_1 x_2 x_3 + d_2 x_2 x_3 x_4 \\
y_3 &= c_3 v_2 + d_3 v_3 && // && = c_3 x_1 x_2 x_3 + d_3 x_2 x_3 x_4
\end{aligned}
$$

Figure 1: Example program with 4 independents and 3 dependents

$$
\begin{aligned}
\nabla v_1 &= x_2 \nabla x_3 + x_3 \nabla x_2 && // && (0, x_3, x_2, 0) \\
\nabla v_2 &= x_1 \nabla v_1 + v_1 \nabla x_1 && // && (v_1, x_1 x_3, x_1 x_2, 0) \\
\nabla v_3 &= x_4 \nabla v_1 + v_1 \nabla x_4 && // && (0, x_4 x_3, x_4 x_2, v_1) \\
\nabla y_1 &= c_1 \nabla v_2 + d_1 \nabla v_3 && // && (c_1 v_1, c_1 x_1 x_3 + d_1 x_4 x_3, c_1 x_1 x_2 + d_1 x_4 x_2, d_1 v_1) \\
\nabla y_2 &= c_2 \nabla v_2 + d_2 \nabla v_3 && // && (c_2 v_1, c_2 x_1 x_3 + d_2 x_4 x_3, c_2 x_1 x_2 + d_2 x_4 x_2, d_2 v_1) \\
\nabla y_3 &= c_3 \nabla v_2 + d_3 \nabla v_3 && // && (c_3 v_1, c_3 x_1 x_3 + d_3 x_4 x_3, c_3 x_1 x_2 + d_3 x_4 x_2, d_3 v_1)
\end{aligned}
$$

Figure 2: Forward mode evaluation of the 4×3 Jacobian

Figure 2. The 4-vector $\nabla v_1 = (0, x_3, x_2, 0)$ has only two nonzero components and can be composed without performing any arithmetic operations. Consequently the calculation of the gradients ∇v_2 and ∇v_3 costs 4 multiplications. Since both of these vectors have three nonzero components each the calculation of the gradients ∇y_i representing the three rows of the Jacobian $\partial y / \partial x$ costs 18 more multiplications and 6 additions. The total count is thus 22 multiplications and 6 additions.

The calculation performed above represents the so-called forward mode of automatic differentiation, with an operations count corresponding to a sparse implementation of this basic procedure. To see how a few operations can be saved by the so-called reverse mode, let us first consider a graphical representation of the dependence between the x_j and the y_i.

In the graph displayed in Figure 3 the nodes represent the independent, intermediate, and dependent variables and the edges represent the elementary partial derivatives between them. In other words, an edge connects v_i and v_j exactly when v_i occurs on the righthand side of the statement that defines v_j on the left. In that case the edge is annotated with the partial derivative of v_j

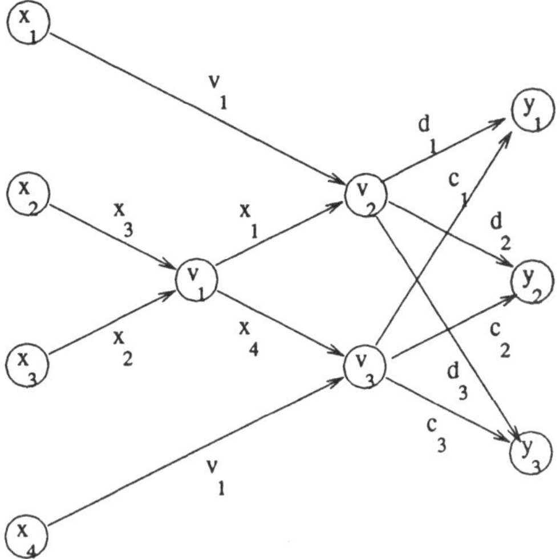

Figure 3: Computational graph with 3 intermediate nodes

with respect to v_i, evaluated at the current argument. Since all operations in our example involve two nonconstant arguments, there are two edges arriving at each node; but in the general situation there are often univariate functions such as exponentials and trigonometric intrinsics, which generate nodes with a single incoming arc. In those situations the calculation of the elementary partials themselves may be nontrivial, though its cost is always bounded by a small multiple of the cost for evaluating the elementary functions themselves. This is true for any realistic measure of computational cost and may account for memory accesses.

Clearly the collection of elementary partials in the computational graph uniquely determines the overall Jacobian. The sparse forward procedure of Figure 2 for calculating the Jacobian combines the elementary partials to arrive at the overall Jacobian by applying the chain rule in one particular way. Contrary to what one might expect, there are several other ways in which the Jacobian can be accumulated from the elementary partials. The results are the same up to round-off, but the computational effort in terms of the number of arithmetic operations may differ.

For example, rather than successively calculating the gradients of intermediates and dependents with respect to the independents, as is done in the forward mode, one can instead compute vectors of sensitivities of the dependents with respect to the intermediate and the independents. More specifically, let us define the *adjoint* vectors

$$\bar{v}_i \quad = \quad (\partial y/\partial v_i) \in \mathbf{R}^3$$

and correspondingly $\bar{x}_i$ and $\bar{y}_i = e_i$. This time the derivative objects associated with the dependents are initialized to Cartesian basis vectors, and we proceed backwards as shown in Figure 4, using the so-called reverse mode of automatic differentiation.

$$
\begin{aligned}
\bar{v}_3 &= d_1\bar{y}_1 + d_2\bar{y}_2 + d_3\bar{y}_3 &&// &&(d_1, d_2, d_3) \\
\bar{v}_2 &= c_1\bar{y}_1 + c_2\bar{y}_2 + c_3\bar{y}_3 &&// &&(c_1, c_2, c_3) \\
\bar{v}_1 &= x_1\bar{v}_2 + x_4\bar{v}_3 &&// &&(x_1c_1 + x_4d_3, x_1c_2 + x_4d_2, x_1c_3 + x_4d_3) \\
\bar{x}_4 &= v_1\bar{v}_3 &&// &&(v_1d_1, v_1d_2, v_1d_3) \\
\bar{x}_3 &= x_2\bar{v}_1 &&// &&(x_2x_1c_1 + x_2x_4d_1, x_2x_1c_2 + x_2x_4d_2, x_2x_1c_3 + x_2x_4d_3) \\
\bar{x}_2 &= x_3\bar{v}_1 &&// &&(x_3x_1c_1 + x_3x_4d_1, x_3x_1c_2 + x_3x_4d_2, x_3x_1c_3 + x_3x_4d_3) \\
\bar{x}_1 &= v_1\bar{v}_2 &&// &&(v_1c_1, v_1c_2, v_1c_3)
\end{aligned}
$$

Figure 4: Reverse mode accumulation of the 4×3 Jacobian

We note that the results are consistent in that the four columns $\triangle x_i$ obtained in the reverse mode form exactly the same matrix as the one formed by the three rows ∇y_i computed earlier in the forward mode. This time, however, the whole procedure requires 18 rather than 22 multiplications and 3 rather than 6 additions. Hence the reverse mode has a lower operations count on this example.

Unfortunately, apart from the computation of gradients, for which the reverse mode is optimal [19], there is no reliable rule to determine a priori whether the forward or reverse mode is better, and there are in fact even more variations on the chain rule. We also note that although it seems as if the reverse mode requires storage that is proportional to the runtime of the undifferentiated code, Griewank [21] has shown that much more reasonable compromises between temporal and spatial complexity can be achieved by a recursive checkpointing approach.

2.2 Variations of Derivative Accumulation

In the "explicit" expressions on the right margins of the informal program listed above we have deliberately used the intermediate values rather than their expansion, for example v_1 instead of x_2x_3. This way one can observe that for each pair (x_j, y_i) of an independent and a dependent variable, the corresponding entry in the Jacobian is a sum over all directed paths connecting them in the computational graph (Figure 3). The additive term associated with each path is simply the product of all edge values involved, which we will call the path product. In our simple example, x_1 and x_4 are connected by only one path to each of the three dependents, whereas x_2 and x_3 impact the same dependents via the intermediates v_2 or v_3.

In general, the number of different paths grows exponentially with the diameter of the graph, that is, the length of the longest directed path in the graph. Therefore, it is in general impossible or at least very expensive to evaluate the additive terms for each path separately and then to add them to the appropriate Jacobian entry. In our example this approach would require 2 multiplications each for the 6 partials with respect to x_1 and x_4 and 3 multiplications each for the 6 partials with respect to x_2 and x_3, yielding a total of 30 multiplications and 6 additions. Both the forward and reverse mode do better because they utilize the fact that certain subproducts like x_3x_1 or x_1d_1 occur repeatedly and can therefore be reused. The forward mode is based on partial path products starting from the independents, and the reverse mode accumulates path products starting from the dependents. In graphs with paths of length four or greater, one can also start forming partial path products in the middle, that is, without involving either independents or dependents initially.

To visualize some particular choices in the usually bewildering variety of accumulating path products, one can think of successively eliminating intermediate nodes by connecting all intermediate predecessor/successor pairs with the products of the corresponding edge values. If a direct edge already exists, its value is incremented by the product of the two arcs running through the intermediate node. In the example above, one could first eliminate the node v_1 by connecting the predecessors x_2 and x_3 with the successors v_2 and v_3 by four edges with the values x_3x_1, x_3x_4, x_2x_1, and x_2x_4. This is essentially what the forward mode does in calculating the gradients ∇v_2 and ∇v_3. Subsequently it eliminates the nodes v_2 and v_3 by multiplying the newly obtained edge values with the constant multipliers c_i and d_i, respectively. The reverse mode first eliminates the node pair (v_2, v_3) and then v_1.

Because of the symmetric role of v_2 and v_3 in Figure 3, there exists only one significantly different third elimination order, namely, v_2, v_1, v_3. Starting from Figure 3, this is depicted in Figure 5, where vw denotes the value of the edge (v, w) and $a+\ =\ b$ is shorthand for $a = a + b$. The thick edges in the graphs on the righthand side are the ones being created or updated in the current elimination step. The total count for this elimination ordering is 23 multiplications and 6 additions, which means it is slightly worse than the forward mode but still a lot better than the explicit evaluation of all path products.

Eliminate node v_2

$$x1y1 = v_1 d_1$$
$$x1y2 = v_1 d_2$$
$$x1y3 = v_1 d_3$$
$$v1y1 = x_1 d_1$$
$$v1y2 = x_1 d_2$$
$$v1y3 = x_1 d_3$$

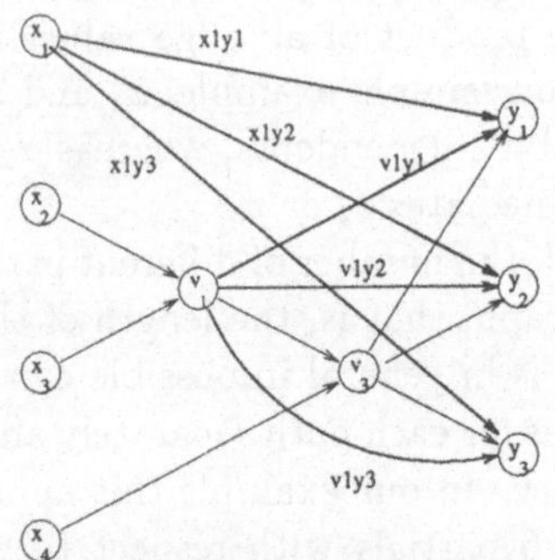

Eliminate node v_1

$$x2y1 = x_3 v1y1$$
$$x2y2 = x_3 v1y2$$
$$x2y3 = x_3 v1y3$$
$$x3y1 = x_2 v1y1$$
$$x3y2 = x_2 v1y2$$
$$x3y3 = x_2 v1y3$$
$$x2v3 = x_3 x_4$$
$$x3v3 = x_2 x_4$$

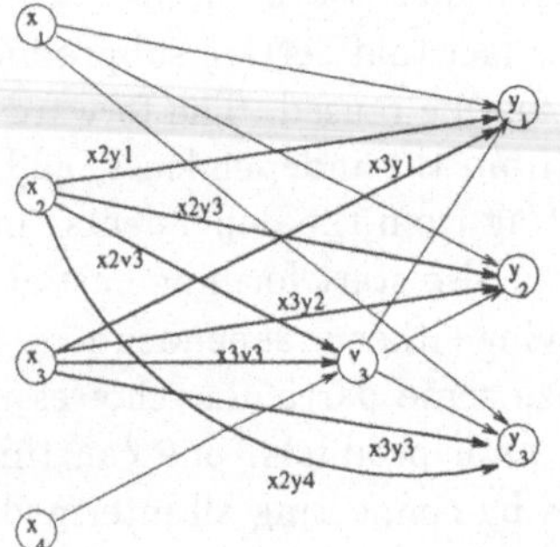

Eliminate node v_3

$$x2y1 \mathrel{+}= c_1 x2v3$$
$$x2y2 \mathrel{+}= c_2 x2v3$$
$$x2y3 \mathrel{+}= c_3 x2v3$$
$$x3y1 \mathrel{+}= c_1 x3v3$$
$$x3y2 \mathrel{+}= c_2 x3v3$$
$$x3y3 \mathrel{+}= c_3 x3v3$$
$$x4y1 = v_1 c_1$$
$$x4y2 = v_1 c_2$$
$$x4y3 = v_1 c_3$$

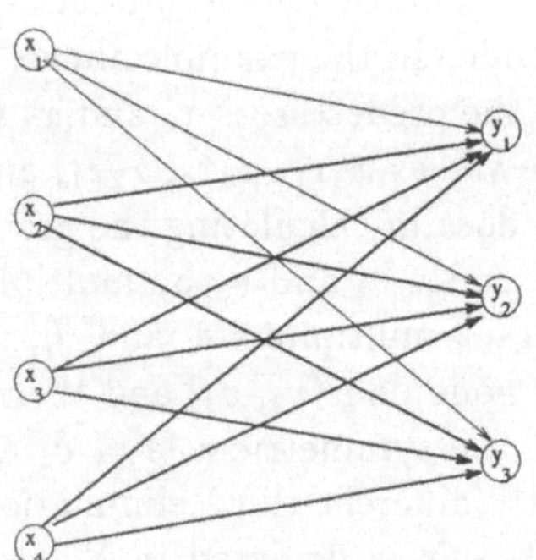

Figure 5: Another possibility for accumulation of the 4×3 Jacobian

The number of multiplications in each elimination step equals exactly the product of the number of predecessors and successors of the vertex being eliminated. For all neighbors this so-called Markowitz degree is changed as some incoming or outgoing edges may merge and others are created because of fill-in. Thus we see that the problem of calculating the Jacobian of a particular vector function using the minimal number of arithmetic operations ([20]) is closely related to the corresponding task in sparse Gaussian elimination. It is therefore not surprising that the Jacobian accumulation problem can be shown to be NP-complete as welll [29]. Just as in the sparse matrix case, the development of near-optimal heuristics for accumulating Jacobians by successively eliminating intermediates provides a rich field for future research.

The classic greedy heuristic is the Markowitz rule, which requires that one always eliminates a vertex of minimal degree, thus minimizing the operations at the current stage. In the long run this may not be the best strategy, as one can already see from the small example given above. Here v_1 has initially the smallest Markowitz degree, namely 4, whereas v_2 and v_3 both have the degree 6. Yet eliminating v_1 first corresponds to the forward mode, which uses more operations than the reverse mode, which eliminates v_3 and v_2 before v_1. While this example is merely of academic interest, there are certain classes of discretized PDEs for which the Markowitz heuristic is a good one [26]. In general, however, these various accumulation strategies may exhibit vastly differing computational complexity. These complexity issues are further explored in [22].

The computational graphs from which the corresponding Jacobians can be obtained by successively eliminating all intermediate vertices contains one node for each arithmetic operation or intrinsic function call. Therefore, the graphs are usually so large that they cannot be stored in core; and even when they can, the interpretive overhead of manipulating them as linked data structures may easily upset the gain in theoretical complexity. Consequently, if one wishes to employ this approach in an AD tool, one must develop a constructive theory of decomposing the graph hierarchically and processing it in pieces. This approach is particularly promising in codes where subroutines with fixed-dimensional inputs (for example finite element stencil evaluators) are called many thousands of times. Since in most applications Jacobians are used only as a means of solving nonlinear systems iteratively, it may be preferable to directly attack the larger problem of computing Newton-steps or their approximations [20], [13].

Like derivatives, many other numerically useful pieces of information about a composite function can be derived from the properties and mutual relations of its elementary constituents. For example, one may determine error bounds, Lipschitz constants, trust region radii, and even parallel evaluation schedules. In this wider sense the term "automatic differentiation" is unduly restrictive, as it does not indicate the ability to build much more general models of nonlinear functions near a given argument. Unfortunately, no convincing alternative terminology has yet emerged.

2.3 Automatic Differentiation Tools

There have been various implementations of automatic differentiation, and an extensive survey can be found in [31]. In particular, we mention GRESS [30] and PADRE-2 [34] for Fortran Programs, and ADOL-C [25] for C and C++ programs. ADOL-C employs the operator overloading capabilities of the C++ language to transparently augment the original code with computations for derivative of arbitrary order. Operator overloading also allows for a flexible and easily customizable system. These issues are further explored in [12].

GRESS, PADRE-2, and ADOL-C implement both the forward and reverse mode. The reverse mode requires that one saves or recomputes all intermediate values that nonlinearly impact the final result, and to this end these tools generate, a trace of the computation. The interpretation overhead of this trace and its potentially very large size can be a serious computational bottleneck [39].

Recently, a "source transformation" approach to automatic differentiation has been explored in the ADIFOR [3] and ODYSSEE [37] projects. Both tools transform Fortran code, applying the rules of automatic differentiation and generating new Fortran code that, when executed, computes derivatives without the overhead associated with "tape interpretation" schemes. ODYSEE generates reverse mode; ADIFOR uses a hybrid forward/reverse mode strategy.

ADIFOR (Automatic Differentiation in Fortran) [3, 7] provides automatic differentiation for programs written in Fortran 77. Given a Fortran subroutine (or collection of subroutines) describing a "function," and an indication of which variables in parameter lists or common blocks correspond to "independent" and "dependent" variables with respect to differentiation, ADIFOR produces portable Fortran 77 code that allows the computation of the derivatives of the dependent variables with respect to the independent ones.

ADIFOR accepts almost all of Fortran 77, in particular arbitrary calling sequences, nested subroutines, common blocks, and equivalences. The ADIFOR-generated code tries to preserve vectorization and parallelism in the original code and employs a consistent subroutine-naming scheme that allows for code tuning, the exploitation of domain-specific knowledge, and the exploitation of vendor-supplied libraries.

ADIFOR employs a hybrid forward/reverse mode approach to generating derivatives. For each assignment statement, it generates code for computing the partial derivatives of the result with respect to the variables on the right-

```
      r$1 = x(1) * x(2)
      r$2 = r$1 * x(3)
      r$3 = r$2 * x(4)
      r$4 = x(5) * x(4)
      r$5 = r$4 * x(3)
      r$1bar = r$5 * x(2)
      r$2bar = r$5 * x(1)
      r$3bar = r$4 * r$1
      r$4bar = x(5) * r$2
  do g$i$ = 1, g$p$
    g$y(g$i$) = r$1bar * g$x(g$i$, 1)
                + r$2bar * g$x(g$i$, 2)
                + r$3bar * g$x(g$i$, 3)
                + r$4bar * g$x(g$i$, 4)
                + r$3 * g$x(g$i$, 5)
  enddo
        y = r$3 * x(5)
```

Reverse Mode for computing

$$\frac{\partial y}{\partial x(i)}, i = 1, \ldots, 5$$

Forward Mode:
Assembling ∇y from $\nabla x(i)$, $i = 1, \ldots, 5.$

Computing function value

Figure 6: Sample of ADIFOR-generated code

hand side using the reverse mode approach, and then employs the forward mode to propagate overall derivatives. For example, the statement

$$y = x(1) * x(2) * x(3) * x(4) * x(5)$$

gets transformed into the code shown in Figure 6. Note that none of the common subexpressions $x(i) * x(j)$ is recomputed in the reverse mode section.

ADIFOR-generated code can be used in various ways [6]: Instead of simply producing code to compute the Jacobian J, ADIFOR produces code to compute $J * S$, where the "seed matrix" S is initialized by the user. So if S is the identity, ADIFOR computes the full Jacobian, and if S is just a vector, ADIFOR computes the product of the Jacobian by a vector. The running time and storage requirements of the ADIFOR-generated code are roughly proportional to the number of columns of S, which equals the gp variable in the sample code above.

3 Exploiting Program Structure and Parallelism

In this section we show how we can exploit high-level program structures and parallelism to speed derivative computations.

3.1 Computing Large, Sparse Jacobians

In the approximation of large, sparse Jacobians bu divided differences, one usually exploits the sparsity structure of these matrices by simultaneously perturbing several "unrelated" input parameters [11, 10]. This structure can also be exploited by a suitable choice of the "seed matrix."

The idea is best understood with an example. Assume that we have a function

$$F = \begin{pmatrix} f_1 \\ f_2 \\ f_3 \\ f_4 \\ f_5 \end{pmatrix} : x \in \mathbf{R}^4 \mapsto y \in \mathbf{R}^5$$

whose Jacobian J has the following structure (symbols denote nonzeros, and zeros are not shown):

$$J = \begin{pmatrix} \bigcirc & & & \\ \bigcirc & & & \diamond \\ & & \triangle & \diamond \\ & & \triangle & \square \\ & & \triangle & \square \end{pmatrix}.$$

So columns 1 and 2, as well as columns 3 and 4 are structurally orthogonal, and in divided-difference approximations one could exploit that by perturbing both x_1 and x_2 in one function evaluation, and both x_3 and x_4 in the other. We can exploit this fact by setting

$$S = \begin{pmatrix} 1 & 0 \\ 1 & 0 \\ 0 & 1 \\ 0 & 1 \end{pmatrix}.$$

For a more realistic example, the 190×190 Jacobian of the blunt body shock-tracking problem described in [38] has only 2582 nonzero entries; its structure is shown in Figure 7a. Because of its sparsity structure, it can be condensed into the "compressed Jacobian" shown in Figure 7b, which has only 28 columns.

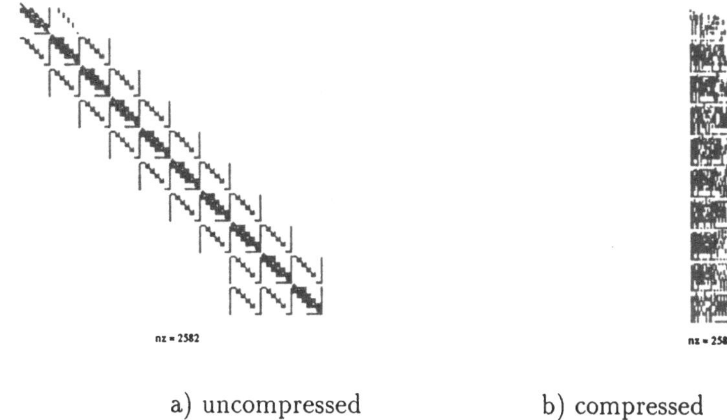

a) uncompressed　　　　　　　b) compressed

Figure 7: Sparsity structure of 190 × 190 blunt body problem Jacobian

	SPARC-2	RS/6000
Sparse DD	0.20	0.130
Compressed ADIFOR	0.13	0.025
Dense ADIFOR	0.85	0.056

Table 1: Performance of ADIFOR-generated derivative code (seconds)

The computation of Jacobian*vector products and compressed Jacobians requires much less time and storage than the generation of the full Jacobian matrix. For example, on the blunt-body problem we observe the performance shown in Table 1 on Sun Microsystems SPARC 2 and IBM RS/6000-550 workstations. The first line gives the run time of a sparse divided-difference approximation, based on the same coloring scheme as the "compressed Jacobian" approach in the second line. The third line shows the running time obtained if one treats this Jacobian as a dense one and ignores sparsity; this means that all derivative operations now are performed with vectors of length 190 instead of 28. As expected, performance suffers, although much less so on the IBM. This is due to the superscalar architecture of this chip and, we suspect, to efficient microcode implementations of multiplications by zero. A comprehensive study of this methodology applied to a collection of optimization model problems can be found in [2].

The compressed Jacobian does not tell the whole story, though. When

	Adds	Mults	Median ∇ length
Compressed	134,428	185,948	28.0
Sparse	5,537	31,633	4.0

Table 2: Dense versus sparse derivative propagation

we keep track of how many operations really have to be performed, avoiding additions and multiplications with zeros, we observe the behavior shown in Table 2. Here we denote the number of additions, multiplications, and the median number of nonzeros in derivative objects for the compressed Jacobian approach and a "sparse" implementation, which avoids operations with zeros. Thus, there is still ample scope for exploiting the temporal sparsity behavior of derivative propagation. This is not surprising since most of the time, the seed matrix is very sparse (e.g., the identity), and derivative objects fill in slowly as the computation proceeds. This effect is most noticeable in the computation of gradients of so-called partially separable functions [6, 8], a rather common class of functions first described by Griewank and Toint [27]. For example, on the 2-D Ginsburg-Landau Superconductivity problem [1], the sparse gradient computation on a 400 × 400 grid requires only 7 percent of the floating-point operations required for the dense version.

The dependency information collected and utilized in the sparse propagation of derivative objects also determines the sparsity pattern of the Jacobian, Hessian, and higher-derivative tensors. A sparse derivative implementation of forward differentiation is currently under development.

3.2 Stripmining Derivative Computations

Let us assume, for example, that we wish to compute a 17-element gradient $g = \frac{dF}{dx(1:17)}$. We can compute any subset of the entries of this gradient through proper choice of the "seed matrix." Hence, we can compute several sets of sensitivities one after the other, or we can decrease turnaround time by spawning several copies of the derivative code, as shown, for example, in Figure 8. Here e_i denotes the i-th canonical unit vector. Hence, the first incarnation of the derivative code will compute the first six entries of the gradient, the second one the next six, and the third one the remaining five.

This "stripmining" approach is simple, much like running several simulations in parallel for divided-difference approximations of derivatives, but has

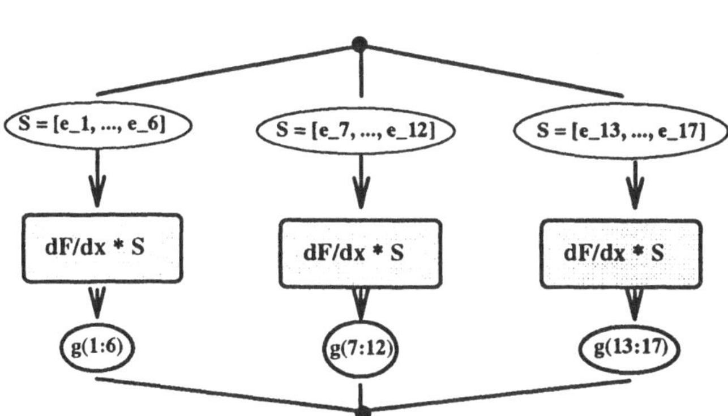

Figure 8: An Example of parallel sensitivity computation

the advantage of decreasing wall clock time with minimal human effort. It can also easily be mapped on any collection of compute nodes, be it a MIMD parallel computer or a heterogeneous workstation network. We implemented this approach with the Fortran M system [16, 15], and the resulting "parallel wrapper" can easily be adapted to other codes.

The TLNS3D code [40] is a high-fidelity aerodynamic computer code that solves the time-dependent, 3-D, thin-layer Navier-Stokes equations with a finite-volume formulation. The code employs grid sequencing, multigrid, and local time stepping to accelerate convergence and efficiently obtain steady-state high Reynolds number turbulent flow solutions. Experiences with ADIFOR on TLNS3D are described in [4, 18].

We were interested in computing the sensitivities of lift, drag, and pitching moment with respect to 60 shape design parameters on a 97x25x17 grid on a workstation platform. Because of memory limitations, we could only fit the code for the computation of up to 12 sensitivities on an IBM RS6000 workstation with 128 MBytes of memory. On the IBM SP1 parallel computer, which for the purposes of this exercise can be regarded as a network of workstations, we observed the performance shown in Table 3. If we had only one processor at our disposal, the memory limitation would limit us to running the 12-sensitivities job five times, requiring an estimated 34,160 seconds, or 9.5 hours. Employing 15 processors, the same set of sensitivities can now be obtained in just about one hour, a dramatic increase in turnaround time. Details are reported in [5].

# Processors	Average Time	Std. Deviation
1	34,160 (est.)	0
5	6,832	53.1
15	3,757	11.5

Table 3: Results of parallel sensitivity computation for TLNS3D

4 Differentiation of Iterative Processes

Until recently it was not clear under what conditions automatic differentiation could be expected to yield useful derivative values if the evaluation program is not just a straight-line code but contains branches dependent on argument values. This situation occurs frequently in programs that employ iterative or adaptive numerical procedures, for example to approximate the solutions of nonlinear algebraic systems or differential equations. One can easily construct examples where branching makes the relation between independent and dependent variables nonsmooth so that the demand for derivative values is unrealistic in the strictly mathematical sense. Nevertheless, one can show under reasonable assumption that on piecewise smooth problems automatic differentiation yields one-sided derivatives, which may be quite useful and informative [32],[14]. More important, Gilbert has shown recently that in the case of contractive fixed-point iterations, the corresponding derivative values also converge to their correct values [17]. We have confirmed the validity of this result on a transsonic fluid dynamics code [4], where the previously used semi-analytic approximation to the derivative of the lift coefficient with respect to the Mach number turned out to be off by 50%. We have extended Gilbert's result to a wider class of iterative schemes, including Broyden's method, and other quasi-Newton schemes.

Our basic assumption is that a parameter dependent nonlinear system

$$F(v, x) = 0 \quad \text{with} \quad F : \mathbb{R}^n \times \mathbb{R}^p \mapsto \mathbb{R}^n$$

is, for fixed x, solved for $v(x)$ by an iteration of the form

$$v_{k+1} \equiv v_k - P_k F(v_k, x) , \tag{1}$$

where P_k is some $n \times n$ matrix, which we will consider as preconditioner. Total derivatives with respect to $x \in \mathbb{R}^p$ will be denoted by primes, and partial derivatives (with v kept constant) by the subscript x.

Simply differentiating (1), one finds that the derivatives of the iterates satisfy the recurrence

$$v'_{k+1} = v'_k - P_k [F_v(v_k, x)v'_k + F_x(v_k, x)] - P'_k F(v_k, x) . \tag{2}$$

Now the question arises under what assumptions this recurrence ensures convergence of the v'_k to the implicit derivative $v'(x)$ defined by the linear system

$$F_v(v(x), x) \, v'(x) \; = \; -F_x(v(x), x) \quad , \tag{3}$$

where $v(x)$ is the implicit function defined by $F(v(x), x) = 0$. In addition to the usual local regularity assumptions on the Jacobian F_v, we imposed the condition that the basic iteration is contractive in that $I - P_k F_v$ has a spectral norm bounded below 1 and furthermore that $P'_k F(v_k, x)$ converges to zero. This last condition is always met if the matrices P'_k are uniformly bounded, as can be expected in the case for Newton's method, where $P_k = F_v(v_k, x)^{-1}$ is continuously differentiable in x provided F is sufficiently smooth. Yet this example also shows that P'_k really should not matter at all since it is multiplied by the residual $F(v_k, x)$, which tends to zero as the iteration progresses. Moreover, in some cases P'_k may not even exist or grow unbounded because of nonsmooth adjustments of the preconditioner P_k from step to step. Then the convergence of the $v'_k(x)$ may be slowed down or prevented altogether. Therefore, it makes sense to *deactivate* the preconditioner P_k, that is, to suppress its dependence on x by simply setting $P'_k = 0$. The resulting simplified derivative recurrence

$$\tilde{v}'_{k+1} \; = \; \tilde{v}'_k - P_k \left[F_v(v_k, x) \, \tilde{v}'_k + F_x(v_k, x) \right] \tag{4}$$

is in general cheaper to perform and corresponds to the so-called incremental iterative schemes proposed in the context of CFD sensitivity analysis [35, 33].

We could show for a large variety of Newton-like methods that it still yields convergence of the $\tilde{v}'_k$ to $v'(x)$ at the same linear rate as the v'_k, which is determined by the spectral radius of $I - P_k F_v$. The only drawback is that the preconditioner P_k must be identified, which might not be so easy in a complex code. Moreover, multigrid schemes and other iterative solvers almost certainly do not satisfy our assumptions, and even if they did, this fact would be very hard to ascertain.

Nevertheless, we have found so far that, for all practical solvers, derivatives of the iterates seem to converge to the correct implicit derivatives with or without deactivation of preconditioners. However, the derivatives tend to lag behind the iterates themselves, and whenever possible one should gauge their convergence by evaluating the derivatives' residual

$$F'(v_k, x) \; = \; [F_v(v_k, x) v'_k + F_x(v_k, x)]$$

which must vanish by the implicit function theorem when $v_k = v(x)$ and $v'_k = v'(x)$. A typical situation is depicted in Fig. 9 for a parameter estimation

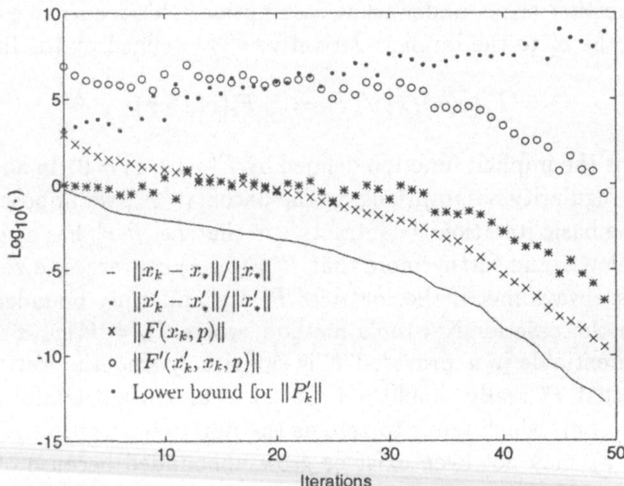

Figure 9: Fully differentiated Broyden's method for Jacobian calculation

problem with three parameters x and a 40-dimensional state vector v. Details are reported in [23].

In Figure 9 it is particularly noteworthy that the P'_k appear to grow unbounded but that the derivative residuals $F'(v_k, x)$ and errors $v'_k - v'(x)$ converge to zero a bit later but essentially at the same rate as the iterates v_k themselves. This seems to be a typical situation.

5 New Approaches in Multidisciplinary Analysis and Design

The task of performing sensitivity analysis on the supersonic or transsonic flow over an elastic wing, for example, combines most of the difficulties that one may encounter in computational differentiation for the purposes of multidisciplinary design analysis and optimization.

1. The two disciplines aerodynamics and structures are mutually dependent, though the coupling is usually assumed to be sufficiently weak that block Gauss-Seydel solvers converge.

2. Both disciplines use their own grids, which usually do not coincide and must in any case be adapted as the calculation proceeds. This requires grid generation and interpolation procedures.

3. The nonlinear systems describing the equilibrium situation are so large and irregular that they cannot be solved directly, but one must employ instead iterative solvers, usually of the multigrid type.

4. The turbulence effects that occur even in subsonic situations must be approximated by algebraic functions, which are typically quite complex and not everywhere differentiable.

Before we discuss how computational differentiation can be made to cope with this kind of situation, let us note that other approaches are even less likely to succeed. Parts of some aerodynamic codes have been differentiated by hand, but since the turbulence models were generally considered too messy for hand manipulation, the results tended to be of very low accuracy. Also, since the outputs of realistic codes are strictly speaking not even continuous functions of the inputs, taking difference quotients is quite dangerous. As we have mentioned before it also has an unnecessarily high computational complexity.

5.1 Solving Coupled Systems

Symbolically we may write the structural and aerodynamic equilibrium conditions as the coupled algebraic system of equations:

$$E(u,v,w,x) \quad = \quad \begin{bmatrix} F(u,v,x) \\ G(u,v,w,x) \\ H(v,w,x) \end{bmatrix} = 0 \quad , \tag{5}$$

where the vectors v and x of linking and design variables may be assumed to have much smaller numbers of components than the vectors u and w representing displacements and flow velocities, respectively. Often the former represent functions in two spatial variables and the latter are discretizations of functions in three spatial variables. Therefore, the overall Jacobian will look roughly like

$$\frac{\partial E(u,v,w,x)}{\partial(u,v,w,x)} \quad = \quad \begin{bmatrix} F_u & F_v & 0 & F_x \\ G_u & G_v & G_w & G_x \\ 0 & H_v & H_w & H_x \end{bmatrix} \tag{6}$$

$$
= \begin{bmatrix}
\times & \times & \times & \times & \times & \times & \times & 0 & 0 & 0 & 0 & \times \\
\times & \times & \times & \times & \times & \times & \times & 0 & 0 & 0 & 0 & \times \\
\times & \times & \times & \times & \times & \times & \times & 0 & 0 & 0 & 0 & \times \\
\times & \times & \times & \times & \times & \times & \times & 0 & 0 & 0 & 0 & \times \\
\times & \times & \times & \times & \times & \times & \times & 0 & 0 & 0 & 0 & \times \\
\times & \times & \times & \times & \times & \times & \times & \times & \times & \times & \times & \times \\
\times & \times & \times & \times & \times & \times & \times & \times & \times & \times & \times & \times \\
0 & 0 & 0 & 0 & 0 & \times & \times & \times & \times & \times & \times & \times \\
0 & 0 & 0 & 0 & 0 & \times & \times & \times & \times & \times & \times & \times \\
0 & 0 & 0 & 0 & 0 & \times & \times & \times & \times & \times & \times & \times \\
0 & 0 & 0 & 0 & 0 & \times & \times & \times & \times & \times & \times & \times
\end{bmatrix} . \tag{7}
$$

We may assume that each discipline has some iterative method for solving the internal system $F(u,v,x) = 0$ and $H(v,w,x) = 0$ for fixed v and x, which amounts to evaluating the implicit functions $u = u(v,x)$ and $w = w(v,x)$. If these methods can be differentiated, then, by the results described in Section 4, one obtains simultaneously approximations to

$$
\left[\frac{\partial u}{\partial v}, \frac{\partial u}{\partial x} \right] = -F_u^{-1}[F_v, F_x] \tag{8}
$$

and

$$
\left[\frac{\partial w}{\partial v}, \frac{\partial w}{\partial x} \right] = -H_w^{-1}[H_v, H_x]. \tag{9}
$$

Note that the inner disciplinary Jacobians F_u and H_w need not be formed explicitly even though their inverses appear in the explicit expression for $\frac{\partial u}{\partial v}$ and $\frac{\partial w}{\partial v}$.

The total derivative with respect to v of the reduced function

$$
\hat{G}(v,x) \equiv G(u(v,x), v, w(v,x), x)
$$

is given by

$$
\frac{\partial \hat{G}}{\partial v} = G_u \frac{\partial u}{\partial v} + G_v + G_w \frac{\partial w}{\partial v} = G' \left[\left(\frac{\partial u}{\partial v} \right)^T, I, \left(\frac{\partial u}{\partial w} \right)^T, 0 \right]^T , \tag{10}
$$

where G' is shorthand for $[G_u, G_v, G_w, G_x]$. Note that we need not compute G' explicitly to evaluate the derivative (10), but rather we can use the "seed matrix" mechanism to compute it by a forward sweep of automatic differentiation at roughly n_v times the cost of evaluating G by itself, where n_v represents the number of linking variables v. By using the resulting Jacobian $\hat{G}_v$ for a Newton-like method to directly solve $\hat{G}(v,x)$ for fixed x, one can directly apply an iterative method based on directional derivatives of $\hat{G}$ in the domain of

v to solve the nonlinear equations (5). In either case the fact that the cross terms F_v, G_u, G_w, H_v are implicitly evaluated through automatic differentiation (rather than simply neglected and assumed to be small) ensures local convergence to a stationary solution, where F_u, H_w, and the overall Jacobian of E with respect to (u, v, w) are nonsingular.

5.2 Design Sensitivities for Coupled Systems

Now suppose we have an objective function $f(u, v, w, x)$ like the lift or the drag coefficient, in which case x incorporates not only geometric and structural parameters but also the free stream boundary conditions. Then the question arises how one can obtain approximations to the total gradient

$$\nabla_x f \quad = \quad f_u \frac{\partial u}{\partial x} + f_v \frac{\partial v}{\partial x} + f_w \frac{\partial w}{\partial x} + f_x \quad . \tag{11}$$

The derivatives $\partial u / \partial x$ and $\partial w / \partial x$ can be obtained by differentiating the individual discipline solvers with respect to x in addition to v. Then one obtains the derivative of $\hat{G}$ with respect to x

$$\frac{\partial \hat{G}}{\partial x} = G_u \frac{\partial u}{\partial x} + G_x + G_w \frac{\partial w}{\partial x} \quad , \tag{12}$$

where we have assumed v to be constant in the differentiation with respect to x. Assuming that $\frac{\partial \hat{G}}{\partial v}$ is small enough to be formed and factored explicitly, one may solve for the root $v = v(x)$ of $\hat{G}(v, x) = 0$ by Newton's method, say, and then evaluate according to the implicit function theorem

$$\frac{\partial v}{\partial x} \quad = \quad - \left(\frac{\partial \hat{G}}{\partial v} \right)^{-1} \frac{\partial \hat{G}}{\partial x} \quad .$$

Even if this is not possible and the iterates v_k are generated instead by an iteration of the form

$$v_{k+1} \quad = \quad v_k - P_k \hat{G}(v_k, x),$$

one may use the simplified derivative recurrence

$$v'_{k+1} = v'_k - P_k \left[\hat{G}_v v'_k + \hat{G}_x \right]$$

with $\hat{G}_v$ and $\hat{G}_x$ as defined above. Provided both levels of the iteration satisfy our contractivity assumption, we can expect that the v'_k will indeed converge to the correct value $\partial v / \partial x$, which can then be used in the expression (11) for the total gradient of the objective function.

6 Outlook

The emergence of robust automatic differentiation tools, together with the increased interest of the scientific community in sensitivity analysis and design optimization has provided a fertile climate for the advancement of computational differentiation techniques. However, despite the ubiquitous nature of derivatives in numerical modelling, computational differentiation techniques as we understand them are still only known or accessible to relatively few computational scientists.

Together with our colleagues in the field, we are working on changing this situation by developing robust, general, and portable tools that provide automatic differentiation technology in a fashion that scales with the problem as well as the computing environment, to support the full scientific development cycle. We are also working on developing templates to guide users in the development of codes embodying certain numerical paradigms to ensure that application of automatic differentiation leads to the desired results.

In the long run, we hope that computational differentiation techniques not only will provide computational scientists with a convenient and efficient means for computing the first- and second order derivatives, but will change the way scientists and algorithm developers view their problems. We expect that the fact that derivatives will be obtainable at a significantly reduced cost and of arbitrary order will spawn new algorithmic approaches as well as new problem-solving environments.

References

[1] Brett Averick, Richard G. Carter, and Jorge J. Moré. The MINPACK-2 test problem collection (preliminary version). Technical Report ANL/MCS–TM–150, Argonne National Laboratory, 1991.

[2] Brett Averick, Jorge Moré, Christian Bischof, Alan Carle, and Andreas Griewank. Computing large sparse Jacobian matrices using automatic differentiation. Preprint MCS–P348–0193, Argonne National Laboratory, 1993.

[3] Christian Bischof, Alan Carle, George Corliss, Andreas Griewank, and Paul Hovland. ADIFOR: Generating derivative codes from Fortran programs. *Scientific Programming*, 1(1):11–29, 1992.

[4] Christian Bischof, George Corliss, Larry Green, Andreas Griewank, Kara Haigler, and Perry Newman. Automatic differentiation of advanced CFD codes for multidisciplinary design. *Journal on Computing Systems in Engineering*, 3(6):625–638, 1992.

[5] Christian Bischof, Larry Green, Kitty Haigler, and Tim Knauff. Parallel calculation of sensitivity derivatives for aircraft design using automatic differentiation. Extended Abstract, submitted to the 5th AIAA/NASA/USAF/ISSMO Symposium on Multidisciplinary Analysis and Optimization, 1994.

[6] Christian Bischof and Paul Hovland. Using ADIFOR to compute dense and sparse Jacobians. ADIFOR Working Note #2, MCS–TM–158, Argonne National Laboratory, 1991.

[7] Christian H. Bischof, Alan Carle, George Corliss, Andreas Griewank, and Paul Hovland. Getting started with ADIFOR. ADIFOR Working Note #9, ANL–MCS-TM-164, Argonne National Laboratory, 1992.

[8] Christian H. Bischof and Moe El-Khadiri. Extending compile-time reverse mode and exploiting partial separability in ADIFOR. ADIFOR Working Note #7, ANL–MCS-TM-163, Argonne National Laboratory, 1992.

[9] Stephen L. Campbell, Edward Moore, and Yangchun Zhong. Utilization of automatic differentiation in control algorithms. To appear in IEEE Trans. Automatic Control, 1993.

[10] T. F. Coleman, B. S. Garbow, and J. J. Moré. Software for estimating sparse Jacobian matrices. *ACM Trans. Math. Software*, 10:329 – 345, 1984.

[11] T. F. Coleman and J. J. Moré. Estimation of sparse Jacobian matrices and graph coloring problems. *SIAM Journal on Numerical Analysis*, 20:187 – 209, 1984.

[12] George Corliss and Andreas Griewank. Operator overloading as an enabling technology for automatic differentiation. Preprint MCS-P358-0493, Argonne National Laboratory, 1993.

[13] Lawrence C. W. Dixon. Use of automatic differentiation for calculating Hessians and Newton steps. In [24], pages 114 – 125, 1991.

[14] Hans-C. Fischer. Differentiation arithmetic and applications in Pascal-XSC. Poster presented at SIAM Workshop on Automatic Differentiation of Algorithms, Breckenridge, Colo., January 1991.

[15] Ian Foster, Robert Olson, and Steven Tuecke. Programming in Fortran M. Technical Report ANL–93/26, Rev. 1, Argonne National Laboratory, October 1993.

[16] Ian T. Foster and K. Mani Chandy. Fortran M: A language for modular parallel programming. Preprint MCS–P327–0992, Argonne National Laboratory, 1992.

[17] Jean-Charles Gilbert. Automatic differentiation and iterative processes. *Optimization Methods and Software*, 1(1):13–22, 1992.

[18] Lawrence Green, Perry Newman, and Kara Haigler. Sensitivity derivatives for advanced CFD algorithm and viscous modeling parameters via automatic differentiation. In *Proceedings of the 11th AIAA Computational Fluid Dynamics Conference, AIAA Paper 93-3321*. American Institute of Aeronautics and Astronautics, 1993.

[19] Andreas Griewank. On automatic differentiation. In *Mathematical Programming: Recent Developments and Applications*, pages 83–108, Amsterdam, 1989. Kluwer Academic Publishers.

[20] Andreas Griewank. Direct calculation of Newton steps without accumulating Jacobians. In T. F. Coleman and Y. Li, editors, *Large-Scale Numerical Optimization*, pages 115–137, Philadelphia, Pa., 1990. SIAM.

[21] Andreas Griewank. Achieving logarithmic growth of temporal and spatial complexity in reverse automatic differentiation. *Optimization Methods & Software*, 1(1):35–54, 1992.

[22] Andreas Griewank. Some bounds on the complexity of gradients, Jacobians, and Hessians. Preprint MCS-P355-0393, Argonne National Laboratory, 1993.

[23] Andreas Griewank, Christian Bischof, George Corliss, Alan Carle, and Karen Williamson. Derivative convergence of iterative equation solvers. Preprint MCS-P333-1192, Argonne National Laboratory, 1992.

[24] Andreas Griewank and George Corliss. *Automatic Differentiation of Algorithms*. SIAM, Philadelphia, Penn., 1991.

[25] Andreas Griewank, David Juedes, and Jay Srinivasan. ADOL-C, a package for the automatic differentiation of algorithms written in C/C++. Preprint MCS-P180-1190, Argonne National Laboratory, 1990.

[26] Andreas Griewank and Shawn Reese. On the calculation of Jacobian matrices by the Markowitz rule. In [24], pages 126 – 135, 1991.

[27] Andreas Griewank and Philippe L. Toint. Partitioned variable metric updates for large structured optimization problems. *Numerische Mathematik*, 39:119–137, 1982.

[28] Uli Häußermann. Automatische Differentiation zur Rekursiven Bestimmung von Partiellen Ableitungen. STUD-102, Institut B für Mechanik, Universität Stuttgart, 1993.

[29] Kieran Herley. On the NP-completeness of optimum accumulation by vertex elimination. Unpublished manuscript, 1993.

[30] Jim E. Horwedel. GRESS: A preprocessor for sensitivity studies on Fortran programs. In [24], pages 243 – 250, 1991.

[31] David Juedes. A Taxonomy of Automatic Differentiation Tools. In [24], pages 315 - 330, 1991.

[32] G. Kedem. Automatic differentiation of computer programs. *ACM Trans. Math. Software*, 6(2):150 - 165, June 1980.

[33] V. M. Korivi, A. C. Taylor, P. A. Newman, G. W. Hou, and H. E. Jones. An incremental strategy for calculating consistent discrete CFD sensitivity derivatives. NASA Technical Memorandum 104207, NASA Langley Research Center, February 1992.

[34] Koichi Kubota. PADRE2, a FORTRAN precompiler yielding error estimates and second derivatives. In [24], pages 251 - 262, 1991.

[35] P. A. Newman, G. J.-W. Hou, H. E. Jones, A. C. Taylor, and V. M. Korivi. Observations on computational methodologies for use in large-scale, gradient-based, multidisciplinary design incorporating advanced CFD codes. NASA Technical Memorandum 104206, NASA Langley Research Center, 1992.

[36] G. M. Ostrovskii, Ju. M. Wolin, and W. W. Borisov. Über die Berechnung von Ableitungen. *Wissenschaftliche Zeitschrift der Technischen Hochschule für Chemie, Leuna-Merseburg*, 13(4):382 - 384, 1971.

[37] Nicole Rostaing, Stephane Dalmas, and Andre Galligo. Automatic differentiation in Odyssee. *Tellus*, 45a(5):558-568, October 1993.

[38] G. R. Shubin, A. B. Stephens, H. M. Glaz, A. B. Wardlaw, and L. B. Hackerman. Steady shock tracking, Newton's method, and the supersonic blunt body problem. *SIAM J. on Sci. and Stat. Computing*, 3(2):127-144, June 1982.

[39] Edgar Soulie. User's experience with Fortran compilers for least squares problems. In [24], pages 297-306, Penn., 1991.

[40] V. N. Vatsa and B. W. Wedan. Development of a multigrid code for 3-D Navier-Stokes equations and its application to a grid-refinement study. *Computers & Fluids*, 18(4):391-403, 1990.

[41] R. E. Wengert. A simple automatic derivative evaluation program. *Comm. ACM*, 7(8):463-464, 1964.

Optimization Problem Formulation for Multidisciplinary Design

Gregory R. Shubin

The Boeing Company, Seattle, WA 98124-0346

1 Summary

In industry today, the design process relies heavily on engineers-in-the-loop, making it is difficult to handle a sufficient number of design variables to optimize a complex product. Additionally, engineering disciplines are usually worked sequentially, so it is tedious and time-consuming to make tradeoffs between disciplines. Multidisciplinary design optimization (MDO) is the formal coupling of two or more computational engineering disciplines with numerical optimization. The promise of MDO is that, by providing engineers with optimization tools and by allowing them to consider several disciplines simultaneously, the current difficulties can be overcome and many benefits can be obtained. Among these benefits are shortening the design cycle, understanding and formalization of design processes and interdisciplinary relationships, and achieving designs that lead to better products at lower cost.

MDO problems offer a richer spectrum of possibilities for overall problem formulation than do single discipline design optimization problems, or multidisciplinary analysis problems. This is because the variables and the equations that characterize the overall MDO problem can be "partitioned" in some interesting ways between what we traditionally think of as the "analysis code(s)" and the "optimizer."

Three fundamental approaches to formulating MDO problems are discussed in the context of aeroelastic (aerodynamics and structures treated simultaneously) wing optimization. The key issue in the three approaches is

	All-at-once (AAO)	Individual Discipline Feasible (IDF)	Multidisciplinary Feasible (MDF)
Use of existing codes	None	Full, no direct coupling of analysis codes	Full, but must couple the analysis codes
Discipline feasibility	None until optimal, then all disciplines feasible	Individual discipline feasibility at each optimization iteration	Multidisciplinary feasibility at each optimization iteration
Variables the optimizer controls. (Thus, these are independent variables in sensitivities.)	Design variables and all analysis discipline unknowns	Design variables and interdisciplinary mapping (coupling) parameters	Design variables
Optimization problem size and sparsity	Very large and sparse	Moderate, size and sparsity dependent on coupling "bandwidth"	Small and dense

Table 1: Comparison of formulation features

the kind of feasibility that is maintained at each optimization iteration. In the most familiar "multidisciplinary feasible" approach, the multidisciplinary analysis problem is solved at each optimization iteration. At the other end of the spectrum is the "all-at-once" approach where none of the individual analysis discipline equations is guaranteed to be satisfied until optimization convergence. In between these extremes lie other "individual discipline feasible" possibilities that amount to maintaining feasibility of the separate analysis disciplines at each optimization iteration, while allowing the coupling between the disciplines to be incorrect until optimization convergence. Some features of these formulations are shown in Table 1, and some speculations on performance of them are given in Table 2.

	All-at-once (AAO)	Individual Discipline Feasible (IDF)	Multidisciplinary Feasible (MDF)
Probable compute time for objective and constraints	Low, evaluate residuals for all disciplines	Moderate, separately analyze each discipline	Very high, full multidisciplinary analysis
Expected overall speed of optimization process	Fast	Medium	Slow
Probability of unanalyzable intermediate designs	Low	Medium	High
Probable robustness	Unknown	High	Medium

Table 2: Comparison of predicted performance

MDO is in its infancy, and many mathematical challenges must be addressed for it to progress. Among these are the efficient computation of optimization gradients, automatic handling of perturbations in complex geometries, development of iterative optimization methods, efficient use of computational parallelism, and understanding of problem posedness.

2   Acknowledgments

The work on problem formulations is joint with Evin J. Cramer and Paul D. Frank of Boeing, and J. E. Dennis, Jr. and Robert Michael Lewis of Rice University. This work is detailed in the References listed below.

3   References

1. Cramer, E.J., Dennis, Jr., J.E., Frank, P.D., Lewis, R.M., and Shubin, G.R. On Alternative Problem Formulations for Multidisciplinary Design Optimization. In Proceedings of the Fourth AIAA/AF/ NASA/OAI
   Symposium on Multidisciplinary Analysis and Optimization, September, 1992, Cleveland, Ohio. AIAA paper 92-4752.

2. Cramer, E.J., Dennis, Jr., J.E., Frank, P.D., Lewis, R.M., and Shubin, G.R. Problem Formulation for Multidisciplinary Optimization. Boeing Computer Services Technical Report BCSTECH-93-026, August, 1993. Also issued as Center for Research on Parallel Computing Technical Report CRPC-TR93334.

Computational Fluid Dynamics in Modelling Blast Furnace Erosion

Alfred Preuer, Günther Kolb, Walter Zulehner

1   Introduction

A blast furnace is a plant where pig iron is produced from ore, see fig. 1. This is the first stage in the liquid metal phase. The lower part of the furnace, the so-called "hearth", forms a reservoir for the iron. The inner side consists of refractory material. It protects the steel shell outside the furnace against the high temperature of iron of about $1500°C$.

During a campaign life of a furnace an erosion process is acting on the refractory wall. A typical erosion profile is given in fig. 2. The maximum erosion occurs at the wall just above the bottom. The erosion at this weakest point restricts the campaign life of the furnace. If this erosion process continues, an uncontrolled break through of iron through the wall is possible. This situation is very dangerous for the operating team working at the furnace and expensive for the company. The left side of fig. 2 shows the erosion profile of a break through.

The aim, therefore, is to analyze the erosion process in order to be able to take appropriate measures for a reduction of the erosion process, which would lead to an increase in security and a reduction of costs by the longer campaign life of the furnace. It is an inverse problem in the sense that the conditions within the furnace are to be found which lead to the observed erosion profiles.

Since no measurements are possible during blast furnace operation, the investigation of the erosion phenomena is performed theoretically by describing the process with an adequate model and by numerical simulation based on this model. Verification is done by comparing the numerical results with water model experiments.

A more detailed description of the problem and a literature survey can be found in [1], [2].

2   The model

From the metallurgy and the experience of the blast furnace engineers one knows that the most important erosion mechanism is chemical erosion by carbon dissolution. Liquid iron contains and tries to dissolve carbon until saturation is reached. Carbon sources in the hearth are coke and the refractory material of the wall. The dissolution of this carbon will be investigated in the following. Necessary for the erosion is that the iron is not saturated with carbon. Then the erosion process can start. Further, a permanent iron flow must exist in the hearth such that the erosion can continue. Hence, the iron flow has to be considered, too. The flow model must include the effects of free convection by hearth cooling and forced convection by tapping off iron through the taphole. Some part of the hearth, the so-called coke zone or "dead man", contains so much coke that the flow in this region has to be modelled as a flow through a porous medium.

2.1 The governing equations

Under the assumptions of constant material properties and using the Boussinesq approximation for free convection and the Ergun approximation for the flow through porous medium the governing equations read:

Continuity equation

$$\frac{\partial u_i}{\partial x_i} = 0, \tag{1}$$

where u_i, $i = 1, 2, 3$ denote the cartesian components of the velocity.

Momentum equations

$$u_j \frac{\partial u_i}{\partial x_j} = -\frac{1}{\rho} \frac{\partial p}{\partial x_i} + \nu \frac{\partial^2 u_i}{\partial x_j^2} + g_i \beta (T - T_{ref}) + S_{u_i} \tag{2}$$

with density ρ, pressure p, kinematic viscosity ν, thermal expansion coefficient β, gravitional acceleration $(g_1, g_2, g_3) = (0, 0, 9.81)$, temperature T, reference temperature T_{ref} and with

$$S_{u_i} = \begin{cases} 0 & \text{in the free flow region} \\ -150 \dfrac{(1 - eps)^2}{eps^3 d^2} \nu u_i & \text{in the coke zone} \end{cases}$$

where eps denotes the porosity and d the particle diameter in the porous medium.

Temperature equation

$$\rho \frac{\partial}{\partial x_i} (u_i T) = \frac{\lambda}{c_p} \frac{\partial^2 T}{\partial x_i^2} \tag{3}$$

with specific heat c_p and thermal conductivity λ, given by

$$\lambda = \begin{cases} \lambda_{Re} & \text{in the free flow region,} \\ eps\lambda_{Re} + (1 - eps)\lambda_{Ko} & \text{in the coke zone,} \end{cases}$$

where λ_{Re} and λ_{Ko} denote the thermal conductivity in iron and in the porous medium, respectively.

Carbon concentration equation

$$\rho u_i \frac{\partial C}{\partial x_i} = \rho D_c \frac{\partial^2 C}{\partial x_i^2} + m_c O_{sp} \tag{4}$$

with carbon concentration C, diffusion coefficient D_c, specific surface O_{sp} in the coke zone and carbon source m_c, given by

$$m_c = K(u, T)\rho \frac{1}{100}(C_{sat} - C),$$

where $K(u, T)$ denotes the mass transfer coefficient depending on velocity and temperature and C_{sat} is the saturated carbon concentration.

The **erosion velocity** $\dot{n}$ is then given by

$$\dot{n} = \frac{m_c}{\rho_c}. \tag{5}$$

2.2 Boundary conditions

The computational domain can be seen in fig. 3.

At the wall of the hearth and at the bottom the following boundary conditions are prescribed:

$$u = 0, \quad \lambda \frac{\partial T}{\partial n} = \alpha_{tot}(T - T_a), \quad C = C_{sat}$$

with heat transfer coefficient α_{tot} and ambient temperature T_a. $\partial/\partial n$ denotes the normal deriviative with respect to the boundary.

The top of the hearth, the so-called bath surface, is assumed to be a circle with radius R with prescribed temperature

$$T = 1550°C.$$

Inside a circle with radius $R/2$ the conditions

$$u_n = 0, \quad \frac{\partial u_\tau}{\partial n} = 0, \quad \frac{\partial C}{\partial n} = 0$$

are set, where u_n and u_τ denote the normal and tangential component of the velocity, respectively. Outside that cirlce inflow conditions

$$u_n = \frac{m}{\rho A}, \quad u_\tau = 0, \quad C = C_{in}$$

are used with mass flow rate m, cross section area A.

At the taphole outflow conditions are prescribed. Finally, the usual symmetry conditions are assumed at the plane of symmetry.

2.3 Numerical computation

The computation was performed with the commercial fluid flow package PHOEN-ICS.

For a fixed set of parameters, such as permeability and shape of the porous medium, boundary conditions, inlet carbon concentration, the problem given by equation (1) $\div$ (5) can be solved ("forward" computation).

As mentioned earlier the actual task is to find conditions within the furnace, such that the computed erosion profile is in agreement with the measured profile.

Therefore, so-called "control parameters" are introduced and varied in a well defined range until agreement between measured and computed erosion profiles is achieved. For each particular parameter setting a "forward" computation must be performed.

Finding out the significant control parameters was part of the model development, they were not known a priori.

3 Control parameters and model verification

The essential parameters for the iron flow are the shape, the position and the permeability of the coke zone, which defines the region of porous medium where the Ergun approximation S_{u_i} is active.

For the erosion the inlet carbon concentration C_{in} is considered as control parameter.

3.1 Determination of the iron flow control parameters

The values of these parameters were determined with the help of water model experiments. It was possible to achieve a similarity between the flow in a blast furnace and in a water model. That means that the Reynolds-, Prandtl- and Graßhof number are of about the same magnitude and the boundary conditions are similar, especially the cooling rate.

The water model serves different purposes: The numerical computation can be verified by comparing numerical with experimental data. In addition, because of the flow similarity one gets some insight into the flow behaviour inside the blast furnace. So the water model can also be used to determine and to adjust the control parameters by some iteration process of experiment and computation.

If this adjustment is finished, the numerical model can be applied to the full scale blast furnace.

Shape and position of the coke zone: A first attempt is given in fig. 4. In this arrangement, a large free flow region exists along the periphery. The experiment and the computation give the same results, compare fig. 4 and fig. 5.

There is no iron flow at the bottom, where the maximum erosion exists. However, a high flow rate is necessary for a high erosion. Therefore, this situation can not be responsible for the erosion process.

A more detailed investigation of the shape of the coke zone was undertaken. In [3] an experimental investigation of the formation and shape of a coke zone in a blast furnace was done, see fig. 6a. Additionally, a similar numerical investigation was performed by the VOEST-ALPINE company, where the pressure equilibrium in the whole furnace was considered. The gas flow behaviour in the upper part of the furnace influences the formation of the coke zone. Fig. 6b illustrates the results. In agreement with the experiment the base of the coke zone is convex shaped, compare fig. 6a with fig. 6b. It is easy to see that, if the coke moves downwards when tapping off iron, the dead man is sitting at the bottom and the free flow region is restricted to a ring-shaped channel along the periphery with a triangular cross section.

This situation was studied in the next step of the determination process. The free flow region was restricted to a ring channel. The comparison of fig. 7 with fig. 8 illustrates again a good agreement between water model experiment and computation. A ring shaped channel flow with a high flow rate is formed in the same region where the high erosion exists. Therefore, the shape and the position of the coke zone is responsible for the restriction of the free flow region to a channel along the periphery.

The permeability: The permeability of the coke was adjusted in a way that special flow behaviour in the channel of the water model and computation was in agreement. See the no flow zone at the bottom of the channel below the taphole as well as the backward flow at the bottom, fig. 7, 8. These effects are influenced by the permeability. If the permeability is decreased, the coke zone becomes more compact, the flow through the coke zone becomes more restricted and the iron takes the way through the channel. In this case no backward flow occurs. The opposite effect can be seen if the permeability is increased.

Application to the blast furnace: By the above-mentioned procedure the iron flow control parameters were adjusted. Now the numerical model can be applied to the blast furnace. Fig. 9 and 10 illustrate the iron flow in the furnace by a floating and a sitting coke zone, respectively. The flow behaviour shows the same effects as the water model experiments.

3.2 Determination of the erosion control parameter

The inlet carbon concentration C_{in} was determined by varying its value until measured and computed erosion velocities were equal. The resulting value of inlet carbon concentration is 4.7 %. Fig. 11 gives the associated carbon concentration in a cross section depending on the carbon sources given by the coke and the refractory wall. The corresponding erosion velocity can be seen in fig. 12b. Compare the good agreement of the measured and the computed erosion profile in fig. 12a, b.

4 Conclusion and industrial application

Starting from a measured erosion profile, a numerical model was developed to simulate the erosion process and to find out the conditions within the furnace causing the erosion. With the help of water model experiments the model was adjusted, verified and then applied to the blast furnace. The formation of a ring shaped channel flow, due to the shape and position of the coke zone, was found as the most important influence for the erosion, see fig. 2 to fig. 12.

Therefore, one must avoid the ring shaped channel flow in order to reduce erosion. For this a series of numerical simulations with different hearth geometries and cooling rates were done and are still in progress. Fig. 12c shows the result of one of these investigations. It was possible to find an arrangement where a significant reduction of erosion was achieved, compare fig. 12b with fig. 12c.

These numerical simulations are part in the planning stage of the new refractory lining for the next blast furnace campaign life starting 1994.

References

[1] Preuer, A.; Winter, J; Hiebler, H.: Computation of the iron flow in the hearth of a blast furnace. Steel Research 63 (1992) No. 4, 139 - 146.

[2] Preuer, A.; Winter, J; Hiebler, H.: Computation of the erosion in the hearth of a blast furnace. Steel Research 63 (1992) No. 4, 147 - 151.

[3] Shibata, K.; Kimura, Y.; Shimizu, M; Inaba, S.: Kinetics of dead-man coke and hot metal flow in a blast furnace. Internal report, Kobe Steel (1990).

Authors' adresses: Dr. Alfred Preuer, Dipl.-Ing. Günther Kolb: VOEST-ALPINE Stahl Linz GmbH, A-4031 Linz, Austria. Doz. Dr. Walter Zulehner, Institut für Mathematik, Johannes Kepler Universität Linz, A-4040 Linz, Austria.

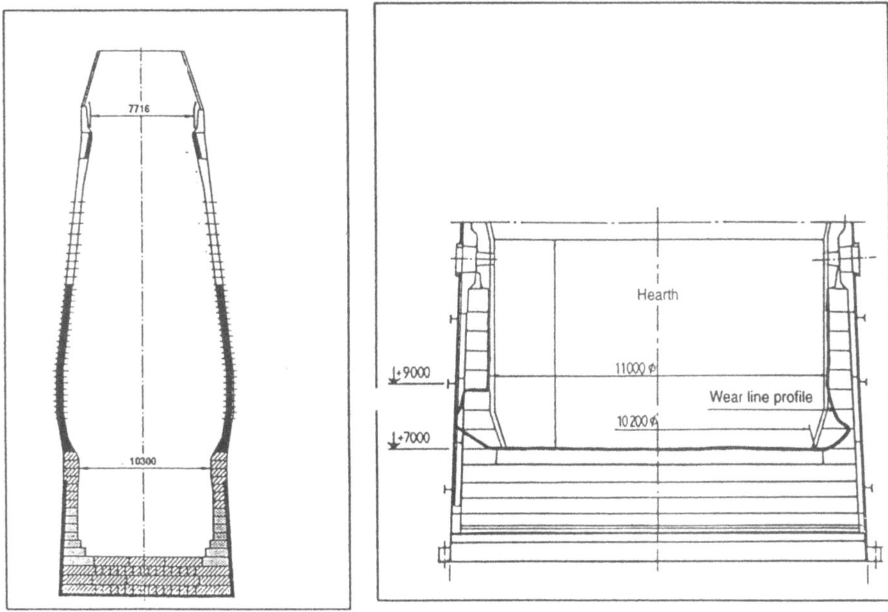

Figure 1 Vertical cross section of a blast furnace

Figure 2 wear line profile of the blast furnace of the VOEST-ALPINE

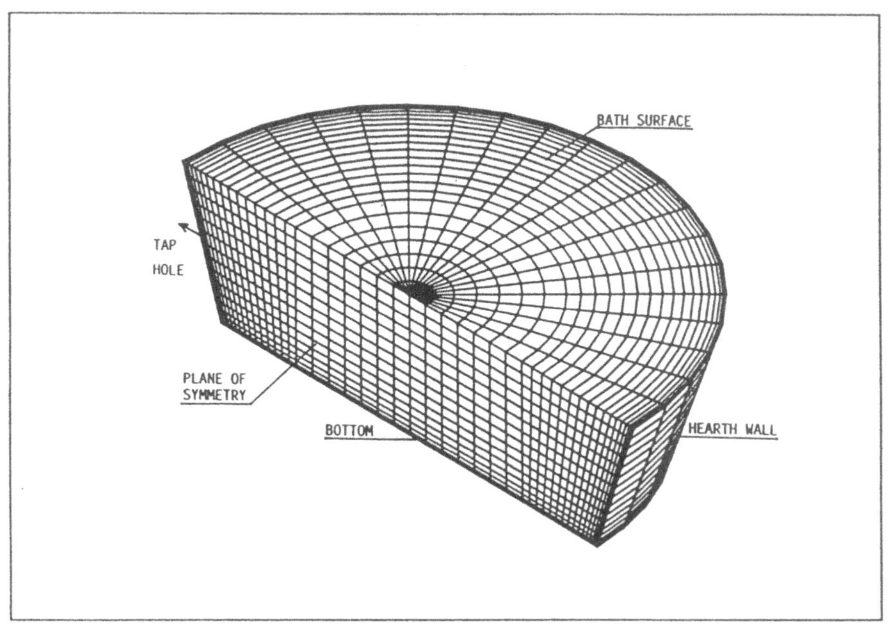

Figure 3 numerical grid of the hearth geometry

Figure 4 Model experiment A, view taphole side, ink addition at the wall 5 cm below the surface; 60 min after ink addition at the bottom (traces are formed within 1 hour) !

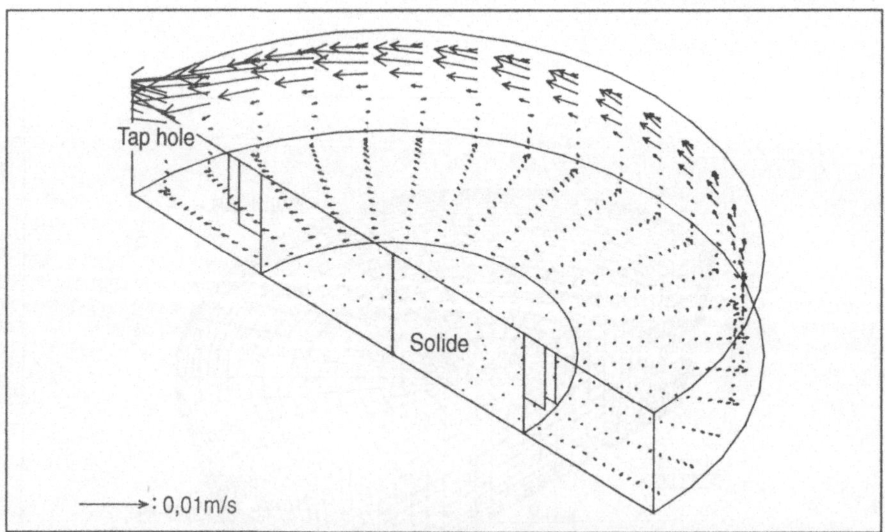

Figure 5 Numerical simulation of model experiment A, velocity distribution 7 cm from the wall and at the bottom

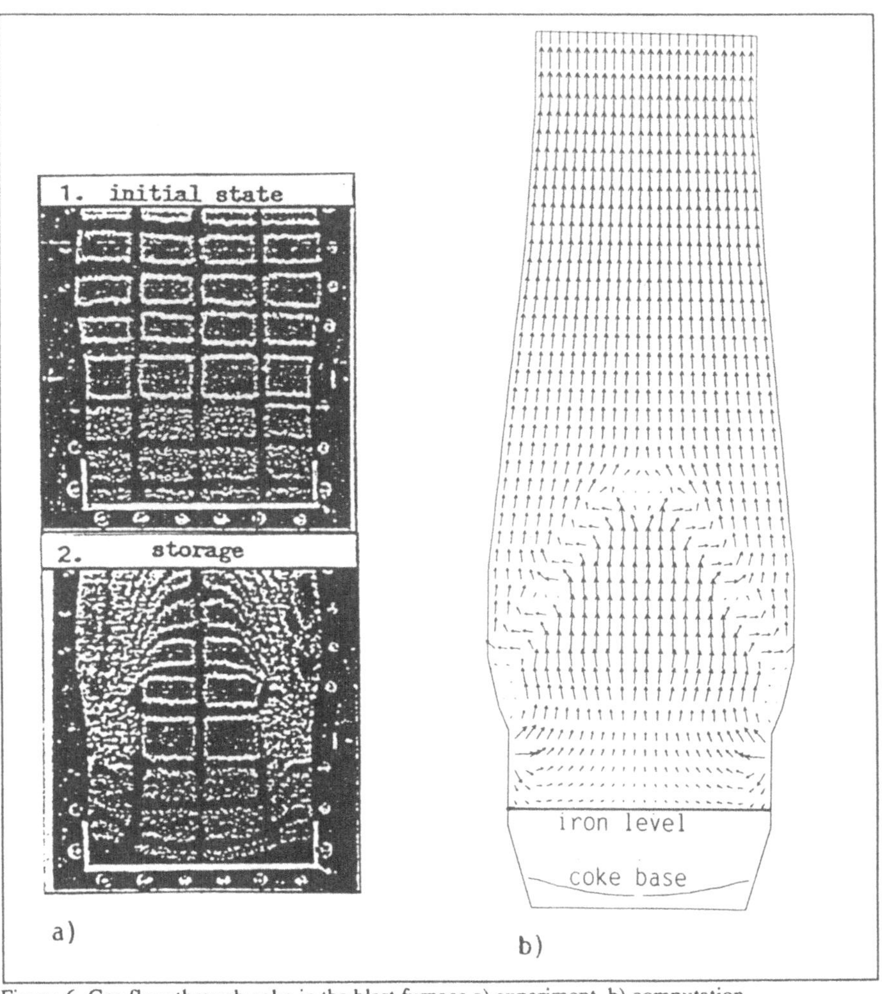

Figure 6 Gas flow through coke in the blast furnace a) experiment, b) computation

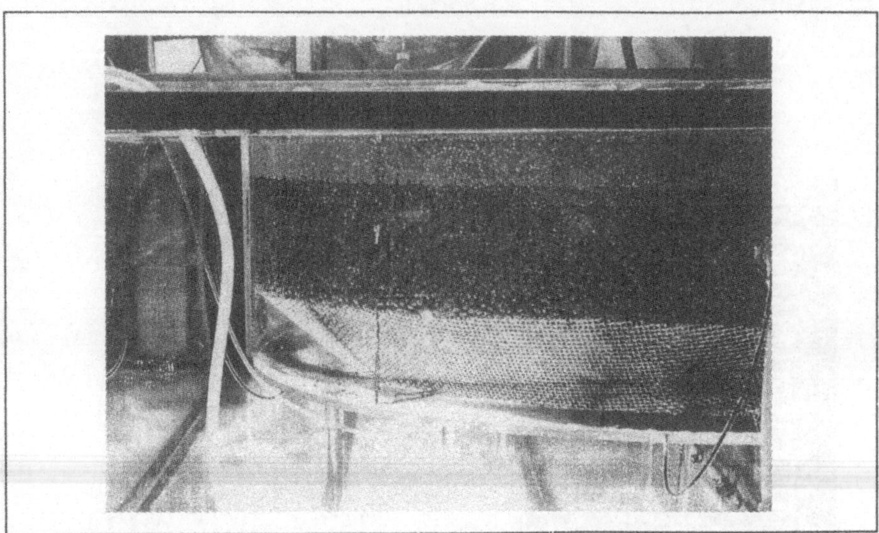

Figure 7 Model experiment B, view taphole side, ring flow in the channel, backward flow at the bottom

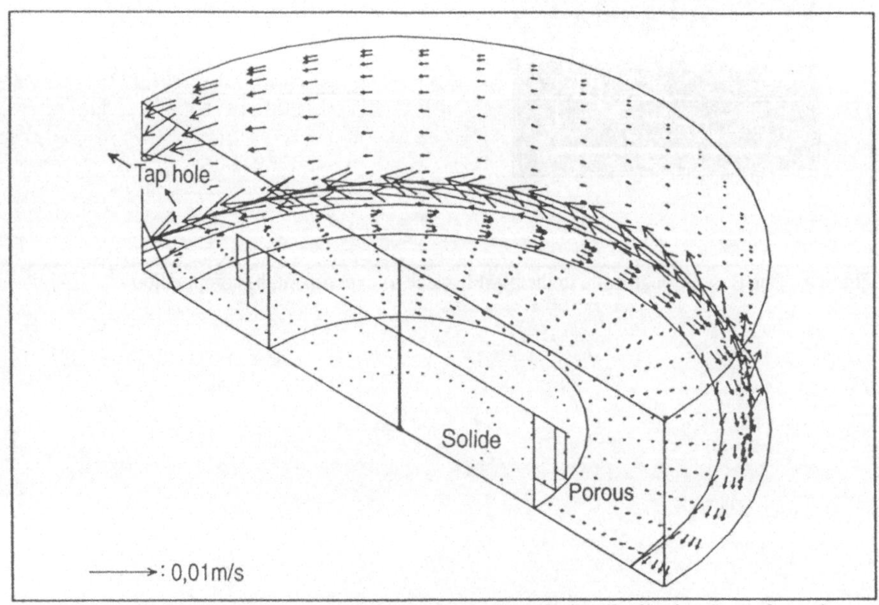

Figure 8 Numerical simulation of model experiment B, velocity distribution 7 cm from the wall and at the bottom

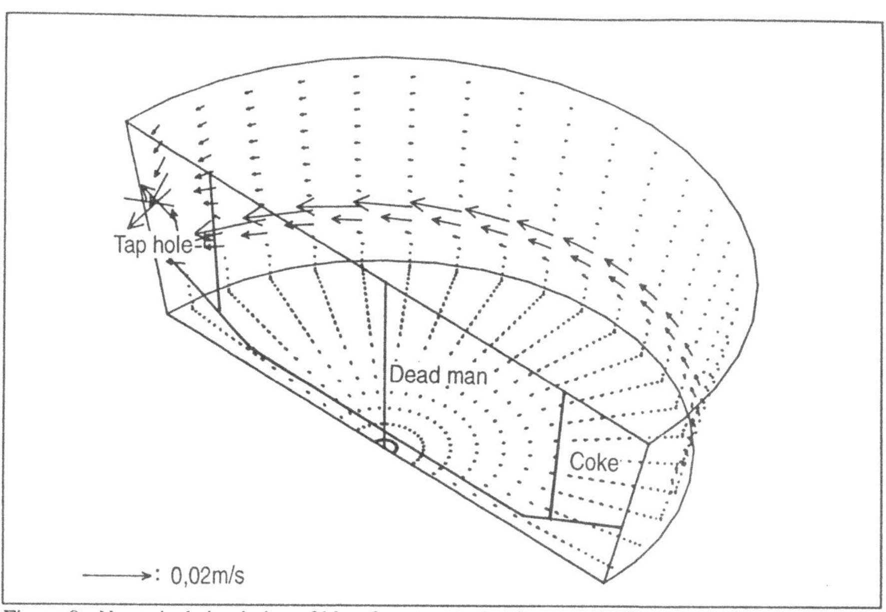

Figure 9   Numerical simulation of blast furnace tapping in case of a floating coke zone (dead man = more compact coke zone)

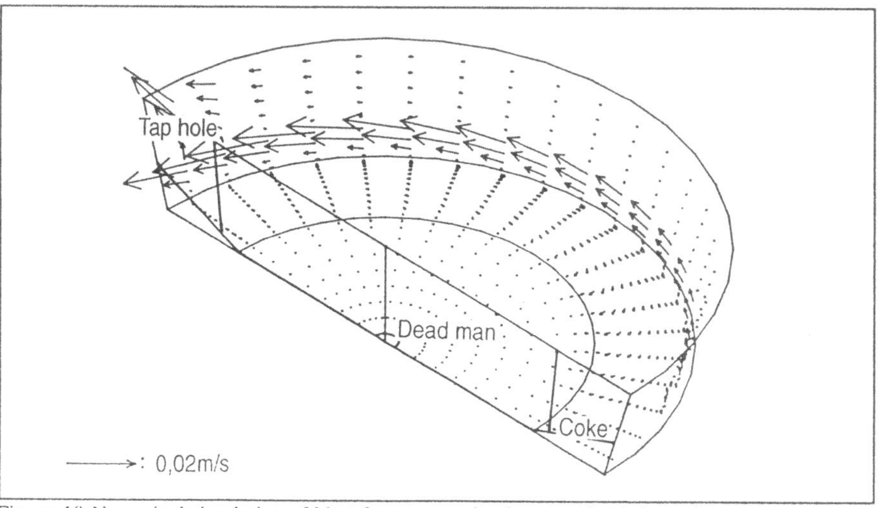

Figure 10 Numerical simulation of blast furnace tapping in case of a sitting coke zone

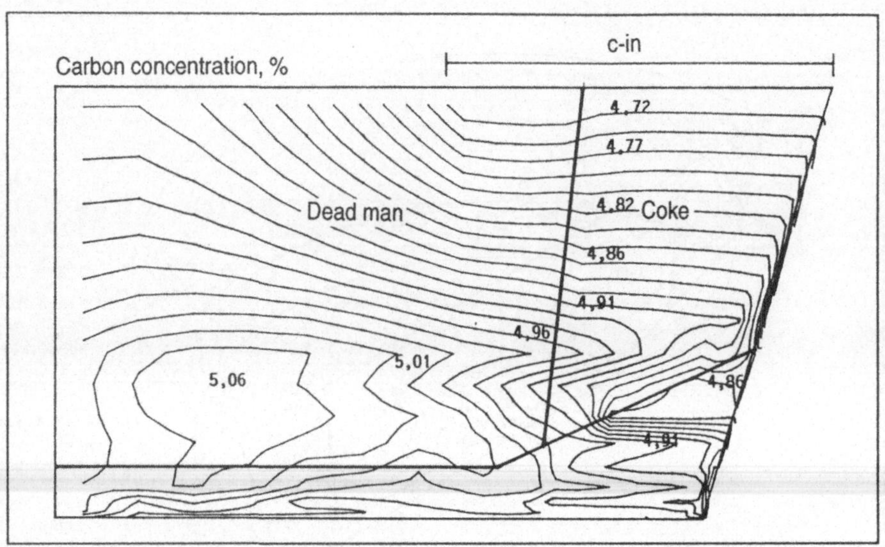

Figure 11 Carbon concentration in the middle plane (= plane normal to the symmetry plane)

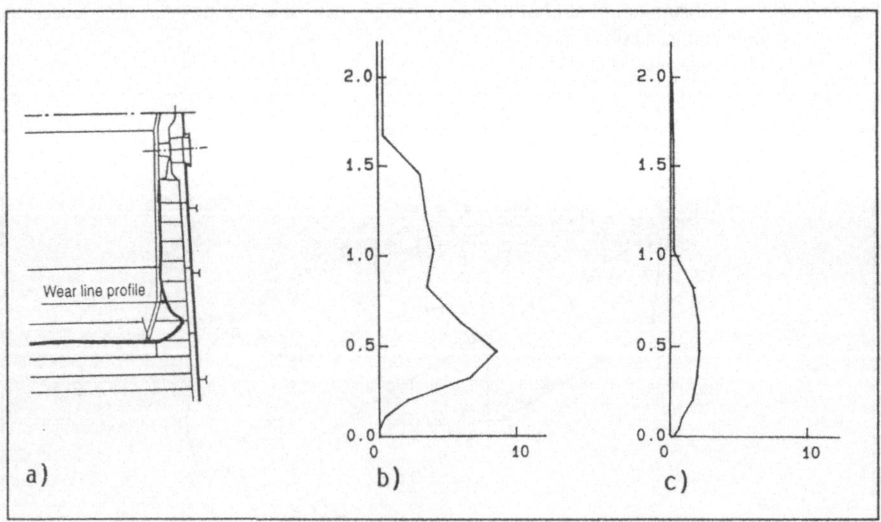

Figure 12 Erosion profile, a) measured, b) computed, c) computed with changed hearth geometry
and cooling rate

A Unidimensional Dynamic Model for the (Ferro-) Silicon Process - A Reliable Tool for Identifying the State of the Furnace ?

Svenn Anton Halvorsen, Elkem a/s

Summary: When producing (ferro)silicon, quartz, (iron ore,) and carbon are charged to an electric smelting furnace, where large amount of energy is supplied. The active part of the furnace is described by a unidimensional dynamic model, where the process is sectioned into an upper shaft zone and a lower hearth zone. The material and heat balances in the shaft are formulated by partial differential equations, while the hearth is described by a system of ordinary differential equations. The unidimensional model has been applied for direct simulations to gain improved process understanding.

Further development will involve use of the model as an identification tool to analyse the state of a furnace. The major problems in this respect are missing experimental data (reaction rates are extremely difficult to measure, etc.) and few relevant and reliable furnace measurements (production rates are for instance only reliable on a weekly average basis). The task will involve not only utilising measurements directly related to the model, but also other measurements and observations where only qualitative correlations with internal furnace states are known.

The success of the present model is due to an interdisciplinary effort. For the further development it will be equally important to merge knowledge within mathematics, control engineering, computer science, chemistry, metallurgy, and practical furnace operation.

1 Introduction

The (ferro)silicon process was developed empirically. Starting in 1892 the French chemist Moissan treated a variety of materials in his electric arc furnace. The combination of quartz, iron ore, and carbon gave a useful product, ferrosilicon, and the process was transferred to a production plant. The scale of the operation has increased steadily, and furnace construction and the auxiliary systems for transport, weighing and control have been modernised, but the basic process is essentially unchanged.

The chemical reactions within the furnace were not investigated for many years, and as late as 1950, they were practically unknown. At this time the management of Elkem started a research program at the Norwegian Institute of Silicate Research with K. Motzfeldt as the main scientist. He proposed a simple stoichiometric model for the process. This model has been further developed and was finally implemented on Excel spreadsheet in 1991 [8,9].

The stoichiometric model has proven valuable for understanding the basic process concepts, and for interpreting various furnace conditions. It does not, however, take into account the dynamic interaction between reaction kinetics, heat and mass transfer, and electrical conduction. Therefore, starting in 1984 a unidimensional dynamic model for the metallurgical aspects (including heat balance) of the process has been developed [2, 3, 4].

The model does only treat pure silicon produced from a charge of pure raw materials (SiO_2 and C). As the silicon chemistry is dominating, they apply, however, to the production of silicon and iron-silicon alloys containing from about 60 to 100% Si.

2 A Unidimensional Dynamic Model

2.1 Brief Process Description - Model Overview

The (ferro)silicon process is normally carried out in large three-phase electric furnaces. A gas filled cavity is formed around each electrode tip, and the final reactions run on the sides and at the bottom of the cavity. The main part of the heat is evolved in electric arcs beneath the electrodes.

The process is cyclic. During operation the cavity increases as the charge materials are consumed. The material layer above the cavity becomes increasingly thinner and hotter. Gradually it becomes white-hot, and smoke-containing gases escape. At last the top layer breaks down or is stoked down and new raw materials are added at the top. Then the cycle is repeated.

In the unidimensional model we intend to study the overall dynamic behaviour of the process, considering a proper description of the upper parts more important than details in the furnace hearth. Hence, the upper region (called the furnace shaft) is described by a coupled set of partial differential equations, while the hearth is modelled by ordinary differential equations. The model should simulate the average behaviour of the active parts of the furnace, i.e. the regions around each electrode. Choosing to stay simple, horizontal heat transfer is neglected, and electric power is supplied at a constant rate to the furnace hearth (supply of carbon from the electrode is neglected). Silicon produced is immediately tapped at the rate of production. Further description is given in [4] and [5].

2.2 Model Equations

The solid components in the shaft are SiO_2, C, SiC, and "condensate", the latter being a mixture of SiO_2 and Si that is formally treated as if it were a separate species, condensed "SiO". Condensate is formed when SiO-gas from the hot part of the furnace is sufficiently cooled, while SiC is made when SiO reacts with carbon [7]. SiO_2, Si, and "SiO" can be present in liquid state. The major gas components are SiO and CO. Other gas species can thermodynamically only be present in negligible amount.

If we assume that the species of each aggregate state move at a common speed in the shaft, the conservation equations for materials and enthalpy can be written as:

$$(1) \qquad \frac{\partial c_i}{\partial t} + \frac{\partial (v_s c_i)}{\partial z} = \sum_j r_{ij} R_j - S_i \quad , \quad i = SiO_2, C, SiC, "SiO"$$

$$(2) \qquad \frac{\partial l_i}{\partial t} + \frac{\partial (v_l l_i)}{\partial z} = \sum_i s_{ij} R_j + S_i \quad , \quad i = SiO_2, "SiO", Si$$

$$(3) \qquad \frac{\partial g_i}{\partial t} + \frac{\partial (v_g g_i)}{\partial z} = \sum_j q_{ij} R_j \quad , \quad i = SiO, CO$$

$$(4) \qquad \frac{\partial H}{\partial t} + \frac{\partial (v_s H_s)}{\partial z} + \frac{\partial (v_l H_l)}{\partial z} + \frac{\partial (v_g H_g)}{\partial z} = \frac{\partial}{\partial z}(\lambda \frac{\partial T}{\partial z})$$

where

t = time (s)
z = vertical location in the shaft (m)
c_i = local concentration of a solid component (moles/m^3)
l_i = local concentration of a liquid component (moles/m^3)
g_i = local concentration of a gas component (moles/m^3)
v_s = downwards velocity for solids (m/s) (negative)
v_l = downwards velocity for liquids (m/s) (negative)
v_g = upwards gas velocity (m/s) (positive)
R_j = reaction rate for reaction no j (moles/s m^3)
S_i = melting rate for component i (moles/s m^3)
r_{ij}, s_{ij}, q_{ij} = stoichiometric coefficient in reaction j for component no i
H = total enthalpy (J/m^3)
H_s = sum of enthalpy for solids (J/m^3)
H_l = sum of enthalpy for liquids (J/m^3)
H_g = sum of enthalpy for gas species (J/m^3)
λ = heat conductivity (W/m K)

Assuming local thermal equilibrium, the temperature can be calculated straight-forward, knowing the local enthalpy and the amounts of each species.

In the furnace hearth solid C and SiC, and liquid SiO_2 and Si can be present, along with the two gas species. For convenience, each 1/2 mole SiO_2 is associated with 1/2 mole Si and formally treated as liquid "SiO". The material and heat conservation equations for the hearth can then be written as

$$(5) \qquad \frac{dM_C}{dt} = -(v_s c_C)_0 - 2\overline{R}_6$$

$$(6) \qquad \frac{dM_{SiC}}{dt} = -(v_s c_{SiC})_0 + \overline{R}_6 - \overline{R}_8$$

$$(7) \qquad \frac{dM_{\text{SiO}}}{dt} = -2(v_s c_{\text{SiO2}})_0 - 2(v_l l_{\text{SiO2}})_0 - (v_s c_{\text{SiO}})_0 - (v_l l_{\text{SiO}})_0 + 2\overline{R}_7$$

$$(8) \qquad \frac{dM_{\text{Si}}}{dt} = (v_s c_{\text{SiO2}})_0 + (v_l l_{\text{SiO2}})_0 - (v_l l_{\text{Si}})_0 + 2\overline{R}_8$$

$$(9) \qquad \frac{dG_{\text{SiO}}}{dt} = -(v_g g_{\text{SiO}})_0 - (\overline{R}_6 + 2\overline{R}_7 + \overline{R}_8)$$

$$(10) \qquad \frac{dG_{\text{CO}}}{dt} = -(v_g g_{\text{CO}})_0 + (\overline{R}_6 + \overline{R}_8)$$

$$(11) \qquad \frac{d\overline{H}}{dt} = W - (v_s H_s)_0 - (v_l H_l)_0 - (v_g H_g)_0 - Q - h_{\text{Si}}\frac{dM_{\text{Si}}}{dt}$$

where

M_i = hearth amount of solid/liquid component no i (moles/m^2)

$\overline{R}_j$ = hearth reaction rate for reaction no j (moles/s m^2)
Tree reactions are involved, the numbering following the definition in [5].

$(...)_0$ = material or enthalpy flux at the shaft/hearth interface

G_i = hearth amount of gas component no i (moles/m^2)

$\overline{H}$ = hearth enthalpy (J/m^2)

W = electric energy input (W/m^2)

Q = radiation heat transfer to the bottom of the shaft

h_{Si} = specific enthalpy for Si at the hearth temperature (J/mole)

The hearth and shaft equations are coupled through the interfacial fluxes. The downward fluxes will be defined by the shaft conditions, while upward fluxes are given by hearth conditions, supplying some boundary conditions for the partial differential equations. Remaining boundary conditions are given by the radiation heat transfer between hearth and shaft, radiation heat transfer at the furnace top, and concentrations and temperature (enthalpy) in the incoming charge mix.

Further details like the chemical equations involved, rate laws for reactions and melting, velocities for the three aggregate states, simplifications by neglecting time derivatives associated with liquid and gas flows, numerical solution by the line method, etc. are discussed in [5].

2.2 Simulations

The unidimensional model has been implemented on an MS-DOS 386 PC with a floating point co-processor. The program can run in two modes regarding the downwards flow of charge materials:

* Continuous feed mode / self-feeding furnace
* Stoking cycles mode

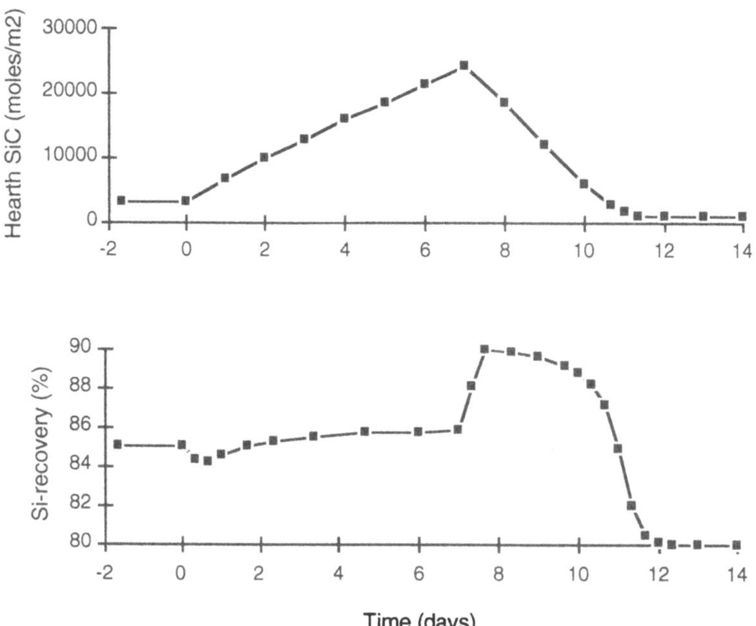

Figure 1 - Dynamic long term behaviour (self-feeding furnace). Stable period followed by one week with a high carbon charge and one week with a low carbon charge.

In the continuous feed mode we assume that voids created by melting or reaction is immediately filled by solid flow from above (refilling continuously with cold charge at the furnace top). This approximation is relevant when simulating long term dynamics (days, weeks) or studying stationary behaviour (averaged over the stoking cycle) [4].

The computed silicon recovery*) for a long term simulation based on the self-feeding approximation is shown in figure 1 (lower graph). Before day zero, the furnace is in a stationary state. Then the molar C/SiO_2-ratio in the charge mix is raised from 1.85 to 1.90. After a transient period the silicon recovery approaches a stationary value of some 86%, which is accompanied by a hearth SiC build-up. Then the charge C/SiO_2-ratio is lowered to 1.80 at day 7, and the furnace moves through a transient period before the recovery settles at 80%. The simulations show that for both transients the initial change in the silicon recovery is opposite to the stationary value.

---

*) The silicon recovery is the amount of silicon produced as liquid metal relative to the amount of silicon charged as SiO_2.

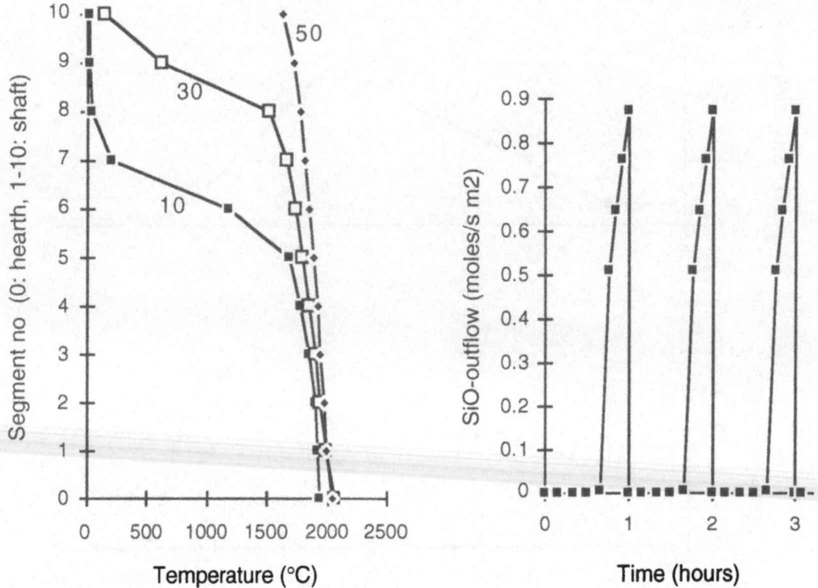

Figure 2 - Typical dynamic short term behaviour for a 60 minutes' stoking cycle. Computed temperature distribution at 10, 30 and 50 minutes after stoking. SiO outflow at the furnace top.

The stoking cycles mode can be used for simulating normal operation (short time dynamics, hours). We let the charge level (shaft top) be constant between stokings. Voids created will in this case cause the shaft to shrink. Stoking is performed at discrete time intervals: Remaining shaft materials are pushed down, partly into the furnace hearth. The shaft is then restored to its full height by adding the necessary amount of fresh charge mix. Further time evolution is then computed by restarting the ordinary time integration.

Some typical computational results are shown in figure 2. Here, the simulation has been continued with constant charge mix until a stationary cyclic behaviour was obtained. The figure shows how the temperature profile in the shaft develops during the stoking period, and the corresponding cyclic loss of silicon as SiO-gas (will burn to microsilica on the furnace top).

Both cases presented are further discussed in [4].

3   Inverse Problems

Most parameters in the model (constants in the reaction rate laws, constants determining the rate(s) of melting, the (effective) shaft height, etc.) are not known, and are further not likely to be experimentally determined in the near future.

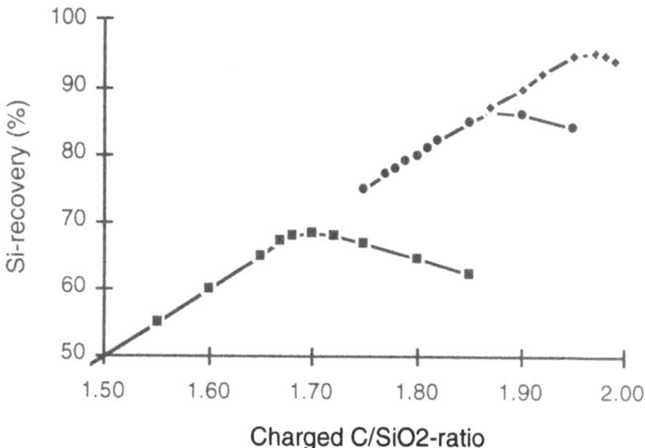

Figure 3 - Typical influence of carbon content in the charge. The stationary silicon recovery is calculated for three different (effective) shaft heights. Squares: 0.5 m, circles: 1.0 m, diamonds: 1.5 m.

Reaction rates are for instance extremely difficult to measure due to the high temperatures and the aggressive atmosphere (SiO-gas). Simulations are nevertheless very valuable as the qualitative process behaviour can be studied, and the relative influence of the various parameters can be found.

We want, however, not only to apply the model for increasing our process understanding, but also as a tool for identifying the state of the process.

The typical influence of the relative amount of carbon in the charge mix can be found by stationary model calculations (running the model in continuous feed mode until a stationary state is established). The silicon recovery will increase (linearly) with increasing carbon content up to a maximum value. Beyond this optimum carbon content, the recovery decreases (almost linearly), figure 3. A true stationary state is only feasible with carbon content below the optimum. With higher carbon content there will be a build-up of SiC in the furnace hearth, which eventually will create poor furnace conditions (narrow hearth with small effective reaction volume). The qualitative influence of the carbon content can also be found by applying a comparatively simple stoichiometric model [9].

We hope to apply the dynamic model both for short term and long term furnace evaluations. On short terms (days) the objective is to adjust the carbon balance (and the setpoints for the electrical control system) to maintain furnace operation close to the optimum point. We want to run the dynamic model parallel to the process, using the model to identify whether the furnace is undercoked (carbon content below optimum) or overcoked (above optimum).

On long terms (weeks, months) the objective is to create conditions that can enable a higher maximum recovery. The location of the optimum carbon content will depend on:

* Raw material properties (reactivity, resistivity, size, ...)
* Furnace conditions (wide/narrow crater zone, effective height and width of shaft zone, particle segregation, even/uneven gas distribution, ...).
* Furnace operation (electrical setpoints, stoking procedures, tapping procedures, ...)

As an example, the qualitative influence of effective shaft height is shown in figure 3.

When a furnace is not running with a sufficiently high maximum silicon recovery, we would like to use the model as a tool for identifying the cause. Further, when some change in raw materials, and/or operating conditions is tested, model simulations should be applied to identify the changes in internal furnace conditions.

Elkem's dynamic model for the (ferro)silicon process has been developed as a direct simulation model. Modifications are necessary in order to apply the model for identification purposes (inverse problems) [11, 12].

A major problem for the identification of inner process states is the lack of precise measurements. The output from the dynamic model relevant for comparing with furnace data, is the amount and temperature of metal produced, and the amount of SiO, CO and the temperature at the furnace top. Due to fluctuating metal accumulation in the furnace, metal weights are only reliable as weekly average values. Temperatures cannot be measured at points corresponding the output from the model, but measurements that can reveal temperature variations in these points are feasible (e.g. variations in the temperature of tapped metal, will indicate variations in the hearth temperature, which is the tapping temperature as seen from the model). The amount of CO produced can be found by off-gas analyses, while SiO will burn to microsilica. The microsilica is collected in baghouse filters. Weights are available with a delay of some two hours (the data being relevant for process evaluation only if there is a separate filter for each furnace). The instantaneous SiO-production can be estimated by measuring the "darkness" in the off-gas (disturbed by other particles that will influence the optical properties, e.g. coal dust) [1].

The measurements described so far seem to be quite insufficient for reliable identifications. We therefore would like to include other measurements and observations. Electrical measurements (currents and voltages) can be analysed and will supply some information on inner process states [11]. Metal analysis and amount and characteristics of tapped slag (mainly unreacted SiO_2) will supply further data. Fluctuations in the amount of various elements in the metal can supply valuable information, even for small amounts that will not (significantly) influence the process behaviour. It is for instance known that the amount of Al and Ca in the metal is correlated to the operating conditions [6]. Manual observations of the furnace top and the tapping conditions also supply valuable information.

Thus Elkem's on-going development on the dynamic model for the (ferro)silicon process involves not only using measurements directly related to the model, but also other measurements and observations where only qualitative correlations are known. Presently it is uncertain how the latter information can be utilised. One option is to set some parameters in the model manually, based on a qualitative evaluation of the available data. Another possibility is to introduce models for how the data is correlated with (internal) conditions in the dynamic model.

4　Concluding Remarks

The (direct) unidimensional model for the (ferro)silicon process has been used for improving our process understanding. Mathematical modelling has been part of a successful research project aiming at improving ferrosilicon production by merging furnace experience with research effort [10]. Further work will include more studies to get additional insight, along with modifications necessary to apply the model as an identification tool.

The contribution from the mathematical disciplines has been essential. It should, however, be stressed that the success has been totally dependent on an interdisciplinary effort. No beneficial results for the metallurgical industry would have resulted without merging knowledge within mathematics, computer science, chemistry, metallurgy, and practical furnace operation.

Further mathematical contribution within an interdisciplinary project is required when modifying the model for inverse problems. Advice will be sought after, both from the theoretical side (mathematics) and from the applied disciplines (control engineering, ...).

5　References

[1]　Gudmundsson, E. B.; Hanneson, T.: On-line measurements of silica dust. INFACON 6, Proceedings of the 6th International Ferroalloys Congress, Cape Town, Volume 1, SAIMM, Johannesburg, 1992, pp 201-204

[2]　Halvorsen, S. A.: Dynamic Model of a Metallurgical Shaft Reactor with Irreversible Kinetics and Moving Lower Boundary, Proc. of the Second European Symposium on Mathematics in Industry, ESMI II, Stuttgart: Teubner 1988, pp 211-223

[3]　Halvorsen, S. A.; Downing, J. H.; Schei, A.: Developing a Unidimensional Simulation Model for Producing Silicon in an Electric Furnace, Proc. of the Fourth European conference on Mathematics in Industry, Stuttgart: B. G. Teubner, Dordrecht: Kluwer Academic Publishers, 1991, pp 293-298

[4]   Halvorsen, S. A.; Schei, A.; Downing, J. H.: A Unidimensional Dynamic Model for the (Ferro)Silicon Process, Electric Furnace Conference Proceedings, ISS/AIME, Vol. 50, 1992, pp 45-59

[5]   Halvorsen, S. A.: Mathematical Models for the (Ferro)Silicon Process, ECMI 93, Proc. of the Seventh European Conference on Mathematics in Industry, Stuttgart: B. G. Teubner, 1994

[6]   Oterkjær, H.: En undersøkelse av Al- og Ca- balansen ved industriell fremstilling av 75% FeSi (in Norwegian, summary in English), Dr.ing. (Ph.D.) thesis, The University of Trondheim, The Norwegian Institute of Technology, 1976

[7]   Schei, A.; Sandberg, O.: Back Reactions during Production of Silicon Metal in a Submerged Arc Electric Furnace, Selected Topics in High Temperature Chemistry, ed. Førland, Grjotheim, Motzfeldt, Urnes, Oslo, Norway: Universitetsforlaget 1966, pp 141-150

[8]   Schei, A.; Larsen, K.: A Stoichiometric Model of the Ferrosilicon Process, Electric Furnace Conference Proceedings, ISS/AIME, Vol. 39, 1981, pp 301-309

[9]   Schei, A.; Halvorsen, S. A.: A Stoichiometric Model of the Silicon Process, Proc. from the Ketil Motzfeldt Symposium, Trondheim, Norway: University of Trondheim, Norwegian Institute of Technology, 1991, pp 41-56

[10]   Tveit, H.; Valderhaug, Aa.: The research project FeSi 8000 - Achieving improvements in ferrosilicon production by merging furnace experience with research effort, Electric Furnace Conference Proceedings, ISS/AIME, Vol. 50, 1992, pp 17-23

[11]   Valderhaug, Aa. M.: Modelling and Control of Submerged-Arc Ferrosilicon Furnaces, Dr.ing. (Ph.D.) thesis, The University of Trondheim, The Norwegian Institute of Technology, 1992

[12]   Valderhaug, Aa.; Balchen J. G.; Halvorsen S. A.: Modelbased Control of the Ferrosilicon Process, paper presented at 3rd IFAC Symposium DYCORD+'92, April 1992, College Park, Maryland, USA.

Address of the author: Svenn Anton Halvorsen
Elkem a/s Research
P.O. Box 40 Vaagsbygd
N-4602 KRISTIANSAND, Norway

ALGORITHMIC ASPECTS AND SUPERCOMPUTING TRENDS IN COMPUTATIONAL ELECTROMAGNETICS

Vijaya Shankar, William F. Hall, Alireza Mohammadian, and Chris Rowell

Rockwell International Science Center, Thousand Oaks, California 91360

Abstract

Accurate and rapid evaluation of radar signature for alternative aircraft/ store configurations would be of substantial benefit in the evolution of integrated designs that meet RCS requirements across the threat spectrum. Finite–volume time domain methods offer the possibility of modeling the whole aircraft, including penetrable regions and stores, at longer wavelengths on today's supercomputers and at typical airborne radar wavelengths on the teraflop computers of tomorrow. To realize this potential, practical means must be developed for the rapid generation of grids on and around the aircraft, and numerical algorithms that maintain high order accuracy on such grids must be constructed.

A structured–grid finite–volume time domain CFD based RCS code has already been developed at the Rockwell Science Center, and this code incorporates modeling techniques for general radar absorbing materials and structures. Using this work as a base, the goal of the CFD based CEM effort is to define, implement, and evaluate various code development issues suitable for rapid prototype signature prediction addressing many issues related to 1) physics of electromagnetics, 2) efficient and higher–order accurate algorithms, 3) boundary condition procedures, 4) geometry and gridding (structured and unstructured), 5) computer architecture (SIMD and MMD), and 6) validation.

Introduction

The design and development of modern aerospace configurations increasingly require better understanding, management and exploitation of ever more complex physical phenomena in different disciplines such as fluid dynamics, structural mechanics, propulsion, controls, and electromagnetics that all play an interdisciplinary role. The traditional approach of serial design by individual disciplines which cannot account for coupling effects will not only result in poor design but also is not cost effective. Extensive simulations of the various complex phenomena are required to understand the interdisciplinary role in design. Along with advances in experimental simulations, with the emergence of the supercomputers, both conventional (vector Cray-like architectures) and massively parallel, the computing technology is beginning to play an ever increasing role in design simulations. Both government and industry view 'Computational Sciences', a discipline that exploits advances in numerical algorithms development and the increasing power of supercomputers, superminicomputers and graphics workstations, as a critical, potentially efficient and cost-effective technology for advanced design.

The development of a computational environment that encompasses different disciplines for multidisciplinary studies as described in Fig. 1 requires that the following computational capabilities be properly matured within each discipline.

1) <u>Computational Fluid Dynamics (CFD)</u>

   — Turbulence modeling
   — Transition
   — Finite-rate chemistry
   — Algorithms for incompressible to hypersonic Mach number range
   — Unsteady and separated flows

2) <u>Computational Electromagnetics (CEM)</u>

   — Proper implementation of Maxwell's equations with general material properties
   — Formulation and implementation of different boundary conditions

3) <u>Computational Structural Mechanics (CSM)</u>

   — Accurate finite-element models for static and dynamic flexible effects
   — Failure and fatigue analysis accounting for material grain structure

The development of computational capabilities in all these disciplines critically depends on the following technologies.

1) <u>Geometry and Grid Setup</u>

   — Three-dimensional geometry modeling
   — Structured and unstructured grid cell representation
   — Pre and post processing for visualization

2) <u>Computer Architectures</u>

   — Vector/parallel coarse grain machines such as the Cray C-90 with 16 CPUs

— Massively parallel machines, both SIMD (Single Instruction Multiple Data) and MIMD (Multiple Instruction Multiple Data)
— Advanced graphics workstations

One of the computational disciplines that has been a supercomputing pace setter for the last three decades is CFD. This technology, which started with the development of the transonic small-disturbance theory in the late 60's, has matured both in algorithm and code development to the point of today being able to solve the time averaged Navier-Stokes equations for predicting the flowfield over a complete fighter. Today, CFD is playing a critical role in the development of next generation fighters and the National Aerospace Plane, though only in single discipline mode. The development of a multidisciplinary computational environment can significantly benefit by applying many of the attributes of CFD to other disciplines such as CEM.

<u>Attributes of CFD</u>

1) The fluid dynamic equations are usually cast in conservation form either as differential or local integral equations conserving mass, momentum, and energy fluxes, thus allowing for numerical capture of flow discontinuities such as shocks and slip surfaces. Equations representing the physics in other disciplines, for example Maxwell's equations in electromagnetics, can be cast in similar conservation forms for conservation of appropiate fluxes.

2) Recent developments of hyperbolic algorithms for solving the time-domain Euler equations are based on the characteristic theory of signal propagation and are referred to as the 'upwind' schemes. For hyperbolic equations, the upwind based schemes can be constructed to provide the right amount of numerical dissipation to achieve both stability and accuracy. Current state of the art in numerical algorithms is based on Essentially Non Oscillatory (ENO) interpolation schemes that can provide arbitrarily high order of accuracy for arbitrary cell shapes such as hexahedral, triangular prism, and tetrahedral elements.

3) For treatment of complex aerospace configurations, CFD methods usually employ either a structured grid based body-fitted coordinate system or an unstructured finite-element grid setup for ease in implementing the various boundary conditions. Similar numerical geometry and grid setup procedures are equally applicable to modeling complex problems in other disciplines.

4) Pre and post processing capabilities running on advanced graphics work stations are effectively employed to visualize and animate the geometry/grid and solution.

The goal of 'Computational Sciences' is to effectively employ the many advances in CFD coupled with emerging supercomputer architectures with expected teraflops (trillion floating point operations per second) performance to mature the computational technology in different disciplines and be able to perform multidisciplinary studies critical to advanced design.

Objectives

Toward establishing a computational environment for performing multidisciplinary studies, the initial goal is to advance the state-of-the-art in CEM with the following specific objectives.

1) Apply algorithmic advances in CFD to solve Maxwell's equations in general form to study scattering (radar cross section), radiation (antenna), and a variety of eletromagnetic environmental (electromagnetic compatibility, shielding, and interference) problems of interest to both the defense and commercial community.

2) Mature the CEM technology to the point of being able to perform coupled CFD/CEM optimization design studies.

3) Establish the viability of MIMD massively parallel architectures for tackling large scale problems not amenable to present day supercomputers.

Computational Electromagnetics (CEM)

The ability to predict radar return from complex structures with layered material media over a wide frequency range (100 MHz to 20 GHz) is a critical technology need for the development of stealth aerospace configurations. Traditionally, radar cross section (RCS) calculations have employed one of two methods: high frequency asymptotics, which treats scattering and diffraction as local phenomena; or solution of an integral equation (in the frequency domain) for radiating sources on (or inside) the scattering body, which couples all parts of the body through a multiple scattering process. A third approach is the direct integration of the differential or integral form of Maxwell's equations in the time-domain.

The time–domain Maxwell's equations represent a more general form than the frequency–domain vector Helmholtz equations, which are usually employed in solving scattering problems. A time–domain approach can, for instance, handle continuous wave (single frequency) as well as a single pulse (broadband frequency) transient response. Frequency–domain–based methods usually provide the RCS response for all angles of incidence at a single frequency, while time–domain–based methods provide solutions for many frequencies from a single transient calculation. Also, in a time–domain approach, one can consider time–varying material properties for treatment of active surfaces. By using Fourier transforms, the time–domain transient solutions can be processed to provide the frequency–domain response. Frequency-dependent (dispersive) and anisotropic material properties can also be included within the time-domain formulation.

CEM is a critical technology in the advancement of future aerospace development through supercomputing. As we transition from the present Gigaflops to the next generation Teraflops computing, CEM will become integral to aerospace design not only as a stand alone technology but also as part of the multidisciplinary coupling that leads to well optimized designs.

CEM Issues

Proper development of a CEM capability appropriate for all aspects of aerospace design must consider various issues associated with electromagnetics. Some of them are:

1) Maxwell's Equations

Maxwell's equations are usually written in a curl form. As mentioned earlier, CFD–based algorithms are usually applied to equations in conservation form, and our current effort applies CFD–based techniques to solve Maxwell's equations. So, the first step in the CEM development is to cast Maxwell's equations in an appropriate conservation form either as differential or local integral equations.

Fluid dynamics equations, such as the Euler or Navier-Stokes equations, in a differential conservation form can be written as

$$Q_t + E_x + F_y + G_z = \text{Source} \qquad (1)$$

where Q is the solution vector and E, F, and G are the fluxes in x, y, and z coordinate directions, respectively. The conservation form readily admits weak solutions such as shock waves.

The integral form of the conservation laws which can easily be derived from the differential form by integrating Eq. (1) with respect to x, y, z over any conservation cell whose volume is V.

$$\int\int\int_V \left(\frac{\partial Q}{\partial t} + \frac{\partial E}{\partial x} + \frac{\partial F}{\partial y} + \frac{\partial G}{\partial z}\right) dx\, dy\, dz$$
$$= \int\int\int_V S\, dx\, dy\, dz = \widetilde{S} \quad . \tag{2}$$

This can be rewritten in vector notation as

$$\frac{\partial}{\partial t}\int\int\int_V Q\, dx\, dy\, dz$$
$$+ \int\int\int_V \left(\vec{\nabla}\cdot\vec{\mathcal{F}}\right) dx\, dy\, dz = \widetilde{S} \quad . \tag{3}$$

In the above,

$$\vec{\mathcal{F}} = E\hat{j} + F\hat{k} + G\hat{l} \quad . \tag{4}$$

Applying the Gauss divergence theorem, we can convert the volume integral into a surface integral.

$$\frac{\partial}{\partial t}\left(\widetilde{Q}V\right) + \int\int_s \left(\vec{\mathcal{F}}\cdot\hat{n}\right) ds = \widetilde{S} \quad . \tag{5}$$

In the above equation, the cell average of the dependent variables are denoted by $\widetilde{Q}$. The outward unit normal at any point of the boundary surface of a cell has been denoted by $\hat{n} = \hat{n}_x\hat{j} + \hat{n}_y\hat{k} + \hat{n}_z\hat{l}$.

$$\widetilde{Q} = \frac{\int\int\int_V Q\, dV}{\int\int\int_V dV} \quad . \tag{6}$$

The integral form of the conservation laws given by Eq. (5) defines a system of equations for the cell average values of the dependent variables.

Maxwell's equations in their vector form are

$$\frac{\partial B}{\partial t} = -\nabla \times \mathcal{E} \tag{7}$$

and

$$\frac{\partial D}{\partial t} = \nabla \times \mathcal{H} - J \quad . \tag{8}$$

The divergence conditions $\nabla \cdot D = \rho$ and $\nabla \cdot B = 0$ are derived directly from Maxwell's equations, where $\nabla \cdot J = -\frac{\partial \rho}{\partial t}$. The vector quantities $\mathcal{E} = (\mathcal{E}_x, \mathcal{E}_y, \mathcal{E}_z)$ and $\mathcal{H} = (\mathcal{H}_x, \mathcal{H}_y, \mathcal{H}_z)$ are the electric and magnetic field intensities, $D = (D_x, D_y, D_z)$ is the electric displacement, $B = (B_x, B_y, B_z)$ is the magnetic induction, and $J = (J_x, J_y, J_z)$ is the current density and ρ is the charge density. The subscripts x, y, z in the vector representation of $\mathcal{E}$, $\mathcal{H}$, B, and D refer to components in respective directions.

In order to apply CFD–based conservation–law form finite–volume methods, Eqs. (7) and (8) are rewritten in the form of Eq. (1),

$$
Q = \left\{ \begin{array}{c} B_x \\ B_y \\ B_z \\ D_x \\ D_y \\ D_z \end{array} \right\} ; \ E = \left\{ \begin{array}{c} 0 \\ -D_z/\epsilon \\ D_y/\epsilon \\ 0 \\ B_z/\mu \\ -B_y/\mu \end{array} \right\} ; \ F = \left\{ \begin{array}{c} D_z/\epsilon \\ 0 \\ -D_x/\epsilon \\ -B_z/\mu \\ 0 \\ B_x/\mu \end{array} \right\} ;
$$

$$
G = \left\{ \begin{array}{c} -D_y/\epsilon \\ D_x/\epsilon \\ 0 \\ B_y/\mu \\ -B_x/\mu \\ 0 \end{array} \right\} ; \ S = \left\{ \begin{array}{c} 0 \\ 0 \\ 0 \\ -J_x \\ -J_y \\ -J_z \end{array} \right\}
\tag{9}
$$

In what follows, the permittivity coefficient ϵ and the permeability coefficient μ are taken to be isotropic, scalar material properties and satisfy the following relationship, $D = \epsilon \mathcal{E}$, $B = \mu \mathcal{H}$. Generalization to tensor ϵ and μ is rather cumbersome but straightforward. The current density J is usually represented by $\sigma \mathcal{E}$, where σ is the material electrical conductivity.

For treatment of complex geometries, a body–fitted coordinate transformation is introduced to aid in the application of boundary conditions.

Under the transformation of coordinates implied by

$$
\tau = t, \ \xi = \xi(t, x, y, z),
$$
$$
\eta = \eta(t, x, y, z), \ \zeta = \zeta(t, x, y, z),
$$

Eq. (9) can be rewritten as

$$
\overline{Q}_\tau + \overline{E}_\xi + \overline{F}_\eta + \overline{G}_\zeta = \overline{S}
\tag{10}
$$

where

$$\overline{Q} = \begin{pmatrix} \frac{\vec{D}}{J} \\ \frac{\vec{B}}{J} \end{pmatrix}, \overline{E} = \begin{pmatrix} \frac{-\vec{\xi} \times \vec{\mathcal{H}}}{J} \\ \frac{\vec{\xi} \times \vec{\mathcal{E}}}{J} \end{pmatrix}, \overline{F} = \begin{pmatrix} \frac{-\vec{\eta} \times \vec{\mathcal{H}}}{J} \\ \frac{\vec{\eta} \times \vec{\mathcal{E}}}{J} \end{pmatrix},$$

$$\overline{G} = \begin{pmatrix} \frac{-\vec{\zeta} \times \vec{\mathcal{H}}}{J} \\ \frac{\vec{\zeta} \times \vec{\mathcal{E}}}{J} \end{pmatrix}, \text{ and } \overline{S} = \begin{pmatrix} \frac{-\vec{J}}{J} \\ 0 \end{pmatrix} \tag{11}$$

where $J = |\partial(\xi, \eta, \zeta)/\partial(x, y, z)|$ is the Jacobian of the transformation and, e.g., $\vec{\xi} = (\partial_x \xi, \partial_y \xi, \partial_z \xi)$. The quantities $\vec{\xi} \times \vec{H}$ and $\vec{\xi} \times \vec{E}$ in Eq. (11) represent tangential magnetic and electric fields at a constant ξ surface. Thus, the fluxes $\overline{E}, \overline{F}$, and $\overline{G}$ are nothing but the tangential fields.

Maxwell's equations can also be cast in integral form as

$$\frac{\partial}{\partial t} \int\!\!\int_V\!\!\int \begin{pmatrix} \vec{B} \\ \vec{D} \end{pmatrix} dV + \int_S\!\!\int \begin{pmatrix} \hat{n} \times \vec{E} \\ -\hat{n} \times \vec{H} \end{pmatrix} dS = 0 \quad, \tag{12}$$

where the six components of $\vec{\mathcal{F}} \cdot \hat{n}$ in Eq. (5) are $(\hat{n} \times E, -\hat{n} \times H)$.

In general, the differential form, Eq. (11), will be employed for finite-volume schemes using a structured grid arrangement, and the integral form, Eq. (12), will be used for unstructured grid cell arrangements using finite-element–like finite–volume schemes.

2) Material Properties

The primary design variables affecting RCS are the shape of the conducting structure and the electric and magnetic polarizabilities of the materials that cover various parts of this structure. Because the illuminating fields are weak, the response of the materials can be taken as linear, so that an effective dielectric permittivity ϵ and magnetic permeability μ can be defined, both of which will in general be complex at the illuminating frequency. The treatment of various limiting material conditions relevant to radar absorbing structures within the framework of time–domain electromagnetics is described below.

Resistance Cards — A thin conducting sheet causes a jump in the tangential magnetic field proportional to the electric current in the sheet, which is given by $\sigma d \left(\hat{n} \times \vec{\mathcal{E}} \right)$, where σ is the electric conductivity of the sheet, d is its thickness, $\hat{n}$ is the local normal to the surface of the sheet, and $\vec{\mathcal{E}}$ is the instantaneous local electric field.

Perfectly Conducting Surface

At typical radar frequencies, the electrical conductivity of metals and other aircraft structural composites is sufficiently high that they can be treated as perfect electrical conductors. In this limit, the electromagnetic fields do not penetrate the conducting surface, and the components of the electric field tangent to the surface vanish at every point. This relation, $\hat{n} \times \vec{\mathcal{E}} = 0$, thus appears as a boundary condition on the solution of Maxwell's equations.

<u>Lossy Materials and Dispersive Media</u> — The imaginary parts of ϵ and μ represent energy absorption. For a time harmonic radar wave at radian frequency ω, their effects are exactly equivalent to adding instantaneous electric and magnetic current conductivities equal to $\omega\, \mathrm{Im}(\epsilon)$ and $\omega\, \mathrm{Im}(\mu)$, respectively, in the Maxwell curl equations:

$$\nabla \times \mathcal{E} = -Re(\mu)\partial\mathcal{H}/\partial t - \omega\, \mathrm{Im}(\mu)\mathcal{H}$$
$$\nabla \times \mathcal{H} = Re(\epsilon)\partial\mathcal{E}/\partial t + \omega\, \mathrm{Im}(\epsilon)\mathcal{E} + \sigma\mathcal{E}$$

For a transient pulse, the frequency dependence of ϵ and μ (dispersive media) must be properly modeled over the bandwidth of the pulse. This leads to an integration over the past history of the fields, but, the value of this integral at every time step can be accurately updated using only the immediately previous values of the field and the integral.

<u>Impedance Boundaries</u> — When the product $\mu\epsilon$ is large compared to $\mu_0\epsilon_0$, the electrical wavelength in the material is correspondingly reduced from its free–space value. To achieve the same accuracy inside a layer of such a material as is needed on the outside, the number of grid points per unit area on the inside surface must be increased by the factor $(\mu\epsilon/\mu_0\epsilon_0)$ compared to the outside surface. In extreme cases, orders of magnitude more grid points would be used in the layer than in all the space surrounding the target. We avoid this problem by eliminating the points inside the layer through the use of an impedance boundary condition applied on its outer surface. The implementation of this condition in the time domain involves an integral over past history that is carried out by the same method that we use for transient–pulse integration with frequency dependent ϵ or μ.

<u>Anisotropic Media</u>

In general, the polarization induced in a material by an applied field need not be parallel to the direction of the applied field. The electrical permittivity ϵ, relating $\vec{D}$ to $\vec{\mathcal{E}}$ and the magnetic permeability μ, relating $\vec{B}$ to $\vec{\mathcal{H}}$ are therefore tensors rather than scalar quantities, and there will typically be different characteristic wave speeds along the three principal axes associated with three tensors. Aside from this formal complication, Maxwell's equations as given in Eqs. (7) and (8) still apply.

<u>Chiral Media</u>

Materials that are predominantly composed of elements of a single parity, such as right–handed helices, exhibit electrical polarization in response to an applied magnetic field and magnetic polarization in response to an applied electric field. At a given frequency, one can define a chiral admittance tensor ξ, in terms of which

$$D = \epsilon E + i\omega\xi B$$
$$H = i\omega\xi E + \mu^{-1} B \,,$$

and the propagation of electromagnetic fields through such a material will again be governed by Eqs. (7) and (8). As in the case of dispersive media, an integration over the past history of the fields will be required to calculate the response to a transient pulse.

<u>Cracks, Gaps, and Wires</u> — Another limiting condition in which resolution requirements become intolerable is for a metal structure having one very small dimension. For instance, putting a grid cell inside a narrow crack can produce excessively large execution times. Fortunately, the local behavior of the fields near these singular geometries can be well approximated analytically. Formulas for the fluxes into reasonably sized grid cells neighboring the singularity can be derived from these asymptotic forms and used to update the local fields.

In order to properly capture the various physics associated with these different material properties, from a numerical implementation point of view, it is important to cast the Maxwell equations in a conservation form given by Eq. (11) or Eq. (12).

3) Boundary Conditions

Proper implementation of various boundary conditions associated with material properties such as perfectly conducting walls, resistive sheets, material interface, impedance boundaries, as well as computational boundary conditions such as nonreflecting outer boundaries are very crucial to accurately modeling problems in electromagnetics. In fact, higher order accurate implementation of boundary conditions in any computational simulation in any discipline is the number one computational issue.

The physical boundary conditions on the electric and magnetic fields at a material interface follow directly from the requirement that Maxwell's equations be satisfied on the interface. In the limiting case of a perfect electrical conductor, these fields vanish inside the conductor, and a sheet of electric current and charge at the interface provides the necessary field discontinuities between the outer and inner surfaces of the conductor.

At the outer limits of the computational domain, the true behavior of the scattered wave is that it propagates outward without reflection. An approximate outgoing–wave condition that can be applied locally at points on the outer boundary will cause some reflection, but errors from this source can be minimized as discussed below.

<u>Perfect Conductors</u> — In a finite–volume scheme, unknown field values are normally computed only at the centroids of the grid cells. However, to capture potentially rapid field variations along a conducting surface, one will have to solve Maxwell's equations right on the conducting surface satisfying the boundary condition on the tangential electric field, $n \times E = 0$, to the same accuracy as any field points (preferably to at least second order accuracy). A rigorous boundary condition implementation procedure based on characteristic theory can be applied to solving Maxwell's equations right on the body points. This step is crucial to capturing the right surface currents accounting for traveling waves. The final RCS results which are obtained from the surface currents are only as accurate as the values one computes for $n \times H$ on the conducting surface.

The boundary condition procedure for perfectly conducting walls can also be appropriately modified and applied for impedance walls, where the surface tangential electric field, instead of being zero, is proportional to the tangential magnetic field.

<u>Outer Boundaries</u> — Along a given direction in space, the local electromagnetic fields can be grouped into forward and backward propagating combinations. One can develop a hierarchy of nonreflecting boundary conditions using the characteristic theory of signal propagation. A simple first order condition imposes the requirement that the incoming scattered field signal normal to the outer boundary is zero. Though this is sufficient for many scattering problems, research in numerical algorithms needs to address the development of higher–order nonreflecting boundary conditions. This will allow one to place the outer boundary very close to the scatter and minimize the number of grid cells in the computational domain.

<u>Material Interface</u> — Across a material interface where the material properties ϵ and μ can be different on either side certain boundary conditions on the tangential fields are to be satisfied. For example, without the presence of

any lossy medium at the interface, the tangential fields $n \times E$ and $n \times H$ are continuous even though the solution vectors D and B will be discontinuous. When a resistive medium is present at the interface, then appropriate jumps in $n \times E$ and $n \times H$ must be accounted for in the boundary condition implementation. In order to compute the right wave reflection and transmission at an interface, the boundary conditions will have to be satisfied to the same level of accuracy as the order of the scheme in the field points. For schemes with order of accuracy greater than two, developing corresponding higher order material interface boundary condition procedures will be quite challenging.

4) Higher–Order Accurate Algorithms

Construction of higher order accurate, stable algorithms that can compute the amplitude and phase of scattered waves accounting for various boundary conditions is very critical for low observable RCS calculations. There are many types of algorithms one can employ to solve the time domain Maxwell equations, namely implicit/explicit schemes, upwind/central difference schemes, single step/multistep schemes, etc. The implementation of higher–order accurate boundary conditions is closely linked to the type of discretization procedure used to solve the Maxwell equations. The choice of algorithm will also depend on whether one uses the differential form or the integral form of the equations. This also determines the choice of gridding, whether it is a structured grid setup or an unstructured cell arrangement. More and more, advanced algorithmic developments in CFD are addressing the use of unstructured grid arrangements and massively parallel computing architectures.

5) Geometry/Gridding

Just like CFD simulations, problems in CEM also involve arbitrarily shaped three–dimensional geometries that need to be represented properly in the computer simulation. In addition to the external shape, CEM also requires modeling the interior of the penetrable structure. The CEM development can greatly benefit from the CFD experience in modeling both the geometry and the grid setup. Depending on the formulation, one may choose either a structured grid or an unstructured grid setup.

Two gridding issues that need to be addressed in EM computations are: 1) number of grid points per wavelength to properly represent the fields in and around a scatterer; and 2) how far should the outer boundary be placed from the scattering object to adequately simulate the nonreflecting boundary condition. In general, the number of points/wavelength is not determined by wavelength alone, and involves the body dimensions (characteristic body size with respect to wavelength) also. The outer boundary location, theoretically, can be right on the body surface itself; however, the computational implementation of nonreflecting boundary conditions requires the outer boundary at a few (2 to 5) wavelengths away from the surface. Again, if one can construct higher order accurate implementations of nonreflecting boundary conditions, the outer boundary can be brought very close to the scattering surface. In general, the necessary grid resolution is provided only around and near the body surface. Between the body and the outer boundary, the mesh is allowed to stretch resulting in very crude (3 to 5 points per wavelength) meshes near the outer boundary regions.

The free space wavelength is reduced to smaller values inside a material (as ϵ and μ become large, the speed of propagation, $c = \frac{1}{\sqrt{\epsilon\mu}}$, goes down, causing the wavelength to scale accordingly). Thus, the grid resolution must take into account material properties to adequately resolve the fields inside material zones.

The number of grid points per wavelength required depends on the order of accuracy of the numerical scheme. A second–order accurate scheme usually requires at least ten grid points per local wavelength. One may be able to use a higher order scheme and minimize the number of grid points. 'However, as the order of accuracy goes up, the scheme will also require more computations per grid point, which may offset the execution savings with fewer grid points.

The requirement that the fields are resolved accurately with proper grid resolution makes CEM problems computationally intensive, requiring large scale supercomputing. For example, to compute the radar cross section of a typical aircraft, say the B2 bomber, at 18 GHz frequency (the wavelength at 1 GHz is 30 cm and at 18 GHz it is 1.667 cm), even if one used 10 grid cells per wavelength, it will require 340 million grid points just on the surface (let alone the gridding requirements for the interior layers) and probably upwards of 10 billion grid points in the computational domain. Such a computational problem will require a teraflops capability. The development of advanced CEM codes will test the limits of massively parallel computing architectures of tomorrow.

6) Computer Architecture

With the emergence of massively parallel computing architectures with potential for teraflops performance, any code development activity must effectively utilize the computer architecture in achieving the proper load balance with minimum internodal data communication. On conventional coarse grain vector/parallel Cray-like machines, one must still employ efficient vector coding architecture and minimize memory conflicts to achieve maximum parallelization.

7) Validation

Once a CEM code is developed, the results must be validated against known exact solutions and carefully tailored experimental data. There are many computational issues such as grid resolution, location of the outer boundary, and accuracy of the boundary condition procedures that can only be addressed through a careful study of many validation cases. The Electromagnetic Code Consortium (EMCC) has a list of validation cases comprising many target shapes specifically designed for validating codes.

Present CEM Capability

At the Rockwell Science Center, we have pioneered the application of computational methods from fluid dynamics to computational electromagnetics. Specifically, finite–volume algorithms for accurate time integration of Maxwell's equations have been constructed using principles developed in Computational Fluid Dynamics over the past two decades. Our progress is summarized in various published reports starting in 1988[1-4].

Because of Rockwell's strong interest in military aircraft design, the most important low observable features have already been incorporated in our CFD–based CEM code, CFDEM. Frequency–dependent permittivities and permeabilities, thin sheets, and impedance boundary conditions are all implemented in the present formulation. The present CEM capability is based on the structured grid formulation employing a finite–volume, upwind Lax–Wendroff scheme applied to the differential form of Maxwell's equations in conservation form.

Some of the salient features of the present Rockwell CFD–based CEM code are

- Time–domain Maxwell's equations

- Proven algorithms from CFD

- Single pulse (multiple frequency, transient) or continuous incident wave (single frequency, time harmonic steady state)

- Numerical grid generation — structured multizone grid

    Complex geometry with layered radar absorbing media

- Lossy or lossless material properties

Frequency and time dependent properties

Thin structures (resistive card, lossy paint)

- Vector/parallel code architecture — 2 GFLOPS demonstrated on the Cray–YMP with 8 processors, and 10 GFLOPS on the Cray–C90 with 16 processors.

  Received the 1990 CRAY Gigaflop Performance Award

  Received the 1993 Computerworld Smithsonian Award

- Pre- and post–processor graphics/animation

- Application to both scattering (RCS) and radiation (antenna) problems

- Ideal for CFD/CEM optimization studies.

The CEM code has been extensively tested for the following geometries.

1) Canonical objects such as spheres, cylinders, ogives, thin rods, cones, airfoils, and a circular disc

2) Almond shaped target

3) Inlets of various shapes (square, circular, curved, $\cdots$) including the presence of infinite ground plane

4) Flat plates of various planforms

5) Double sphere

6) Complete wing geometries with layers

7) Finned projectile and cone–cylinder combinations

8) Scattering from ship–like targets

Some sample results are shown here to illustrate the present capability.

Figure 2 shows RCS validation results for a number of flat plate geometries designed and experimentally tested by EMCC solely for the purpose of code validation. Figures 3 and 4 show similar code validation results for a square inlet for both vertical and horizontal polarization. Figure 5 shows the scattering electromagnetic fields around the complete X–31 fighter. Figure 6 shows the hardware performance monitor output using the Atexpert system on the C–90. On a 16 CPU C–90, each CPU is performing at a speed of 650 MFLOPS. Using the autotasking mode, the code achieves an effective usage of over 15 CPUs with a good load balance through efficient parallelization features built into the code.

<u>Plans for the Structured Grid CEM Code</u>

The structured grid CEM code, though it has gone through a considerable development phase, still needs improvements in the following areas:

1) Treatment of singular regions near sharp edges, cracks, and thin wires

2) Accurate prediction of traveling waves on complex bodies

3) Zonal boundaries with a patch option to prescribe many different boundary conditions (the current version only allows one form of a boundary condition at a zonal boundary)

4) Incorporation of anisotropic and chiral media

5) Higher order accurate nonreflecting outer boundary conditions to allow bringing the outer boundary closer to the object

6) Adaptation of the structured grid code for a massively parallel machine.

Planned CEM Development

The objectives of this planned CEM development are to:

1) Develop an unstructured grid cell based CEM code using higher–order polynomial basis functions[5] representing the electric and magnetic field variation in each cell. The polynomial reconstruction procedure will employ both the cell averaged centroidal field values as well as the Riemann fluxes at cell interfaces. Such an unstructured grid based CEM code will allow modeling of very complex scattering objects with interior layers through ease in setting up the unstructured grid around any arbitrarily shaped topologies. The present CEM code, which employs a structured multizone grid arrangement, is rather cumbersome to use in modeling complex shapes.

2) Adapt the unstructured CEM code to massively parallel MIMD computing architectures. The future of supercomputing will increasingly employ such parallel architectures as we approach teraflops performance. Not only CEM will require teraflop computing; as we integrate CEM with other technologies for multidisciplinary computations, the total computational requirement will become very significant.

3) Develop pilot computational capabilities for CFD/CEM optimization studies.

CEM Applications

The primary objective of this CEM development is to advance the state of the art in low observables stealth technology. However, the CEM technology is so generic that it has a wide range of applications that can benefit not only defense programs but also many commercial developments. Some of the applications of this CEM technology are:

1) RCS Prediction

The CEM technology which solves the time–domain Maxwell's equations is well suited for computing the radar return from very complex low observable platforms with radar absorbing layers. An example of the use of CEM technology to predict the RCS of low observable kinetic energy penetrators is shown in Fig. 9. With supercomputers showing potential for teraflops speeds with tens of gigawords of memory, the CEM technology will soon be able to handle complete low observable platforms over a wide frequency range (100 MHz to 20 GHz).

2) CFD/CEM/CSM Coupling

The CEM technology is one of the critical tools required for performing mutidisciplinary coupling studies. Design of advanced low observable platforms will require many tradeoff studies to achieve the various constraints imposed by different disciplines. Timely maturation of the CEM technology addressing the algorithms, physics, validation, and computer architecture issues will benefit the development of a computational environment for multidisciplinary studies shown in Fig. 1.

3) Antenna and Printed Microstrip Circuit Modeling

The CEM technology is applicable not only for computing scattering problems encountered in RCS studies but also can effectively solve electromagnetic problems involving radiation such as those in antenna modeling, and propagation of pulses in printed microstrip circuits. The design of phased array antennas using either slotted waveguides or microstrip patches can be effectively analyzed using CEM. Figure 10 shows some examples of the use of CEM for such applications.

4) Electromagnetic Rail Gun Launch Dynamics

Multidisciplinary coupling problems involving fluid dynamics, electromagnetics, and structures similar to the ones in aerospace are encountered in other areas also. One such area is the design of electromagnetic rail gun launchers for hypervelocity projectiles. Because of the extreme conditions encountered

in hypervelocity electromagnetic launch, many phenomena normally ignored
in conventional ballistics become important for projectiles fired from rail guns.
The dynamics of the projectile package during launch, and of the various elements of the package as they separate during the launch transition phase,
cannot be adequately modeled without accounting for both electromagnetic
and aerodynamic coupling among the parts, as well as deformation and flexure
of each component. Solving the time–dependent form of Maxwell's equations
is an absolute requirement for carrying out accurate simulation of the launch
process.

5) Detection and Identification of Hidden Targets

The problem of detection and identification of buried objects has a wide
range of applications with both military and civilian significance. Detection
and identification of antitank and antipersonnel carriers, underground pipes,
buried waste, oil deposits, and coal–seams are just a few of these applications.
The CEM technology coupled with inverse scattering methods can be a powerful tool for such applications.

6) EM Compatibility/EM Interference/EM Pulse

The response of a system to a broad range of frequencies, or to a sharp
pulse of illumination, is best computed by the time integration of Maxwell's
equations to determine the history of the electric and magnetic field variations on the structure. The CEM codes can effectively model a variety of
electromagnetic environmental effects (E^3), such as field penetration and coupling, interference among different radiating systems on the same platform,
and modification of far–field patterns by near–zone structures. An example
EMP problem computed by the CFDEM code to predict the electromagnetic
pulse response at all locations on the aft section of a Navy ship is shown in
Fig. 11.

7) Biomedical Applications

The CEM technology has many applications in the biomedical area. Some
of them are microwave heating of cancer tissues (hyperthermia treatment) and
magnetic imaging of interior organs such as brain scanning (magnetic resonance
imaging).

References

1. V. Shankar, W.F. Hall, and A.H. Mohammadian, Proc. IEEE 77 (1989),
   709.

2. V. Shankar, AIAA Paper No. 89-1987, AIAA 9th Computational Fluid Dynamics Conference, Buffalo, 13-15 June 1989.

3. A.H. Mohammadian, V. Shankar, and W.F. Hall, Computer Phys. Comm. 68 (1991), 175-196.

4. V. Shankar, AIAA Paper No. 91-1579, AIAA 10th Computational Fluid Dynamics Conference, Honolulu, 24-27 June 1991.

5. A. Harten and S.R. Chakravarthy, UCLA Computational and Applied Mathematics (CAM) Report 91-16, September 1991; also ICASE Report 91-76, September 1991.

6. K.Y. Szema, et al., AIAA Paper No. 92-0150, 30th Aerospace Sciences Meeting, January 6-9, 1992, Reno, Nevada.

7. S.R. Chakravarthy and K.Y. Szema, Technical Report – Phase I, Contract No. N60530-90-C-0393, Naval Weapons Center, China Lake, California, November 1991.

8. S.R. Chakravarthy and S. Palaniswamy, in Proceedings of the Scalable High Performance Computing Conference (SHPCC-92), IEEE Computer Society, Williamsburg, VA, April 26-29, 1992.

Figure and Table Captions

Fig. 1. Integration of multidisciplinary technologies–role of Computational Science in aerospace design.

Fig. 2. Monostatic RCS for HH–polarization, computed by CFDEM (diamonds) and measured (solid line) for different flat plate geometries.

Fig. 3. Monostatic RCS for a square inlet. Horizontal (HH) polarization.

Fig. 4. Monostatic RCS for a square inlet. Vertical (VV) polarization.

Fig. 5. RCS computation for the complete X–31 fighter using the CFDEM code.

Fig. 6. Typical 'Atexpert' hardware performance monitor output from the Cray–C90 showing 10 gigaflop performance for the CFDEM code.

Fig. 7. Unstructured surface grid setup for a fighter with stores.

Fig. 8. Unstructured grid–based CFD solution showing surface pressures for a complex fighter.

Fig. 9. RCS studies for a finned projectile using the CFDEM code.

Fig. 10. CEM applications for antenna and printed microstrip circuits.

Fig. 11. Electromagnetic field computations over the aft section of a ship target for a triangular sweeping pulse. Both stern–on and portside–to incidence are modeled.

Table 1. Demonstration of scalability of CEM algorithms on an MIMD massively parallel computer.

ELECTRO MAGNETICS
CONTROLS
STRUCTURE
ADVANCED DESIGN
PROPULSION
CFD

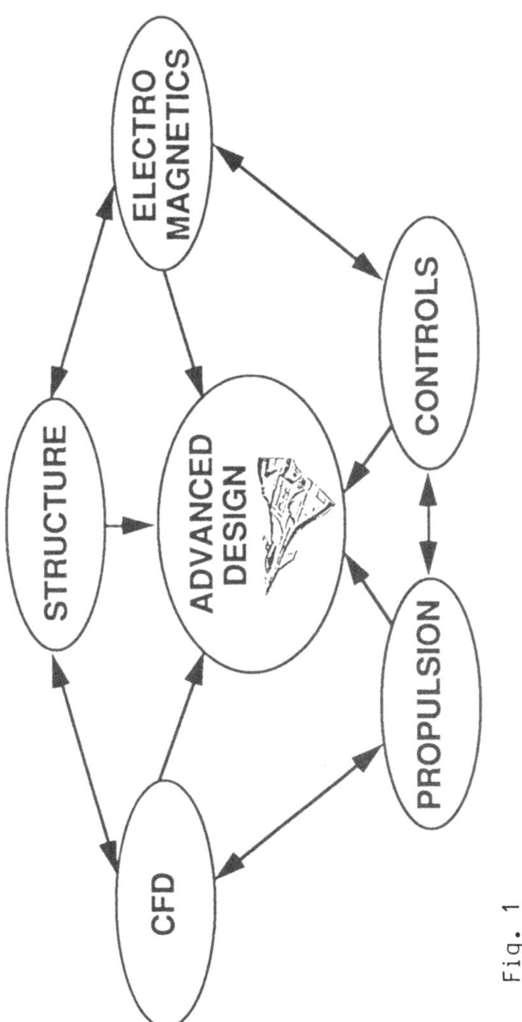

Fig. 1

H-H MONOSTATIC RCS - EMCC Target2
"Target2_hh_CEM"
"Target2_hh_DATA"
RCS dBsm
20
10
0
-10
-20
-30
-40
-50
-60
0 20 40 60 80 100 120 140 160 180
Azimuth
1.0 λ₀ RADIUS
60°
2.0 λ₀
1.0 λ₀

H-H MONOSTATIC RCS - EMCC Target1
"Target1_hh_CEM"
"Target1_hh_DATA"
RCS dBsm
20
10
0
-10
-20
-30
-40
-50
-60
0 20 40 60 80 100 120 140 160 180
Azimuth
1.0 λ₀ RADIUS
60°
2.0 λ₀

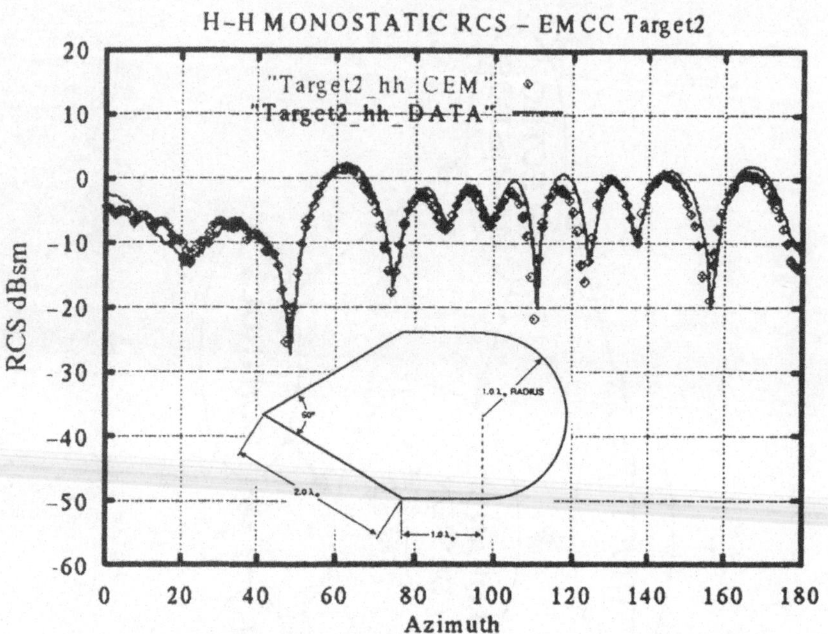

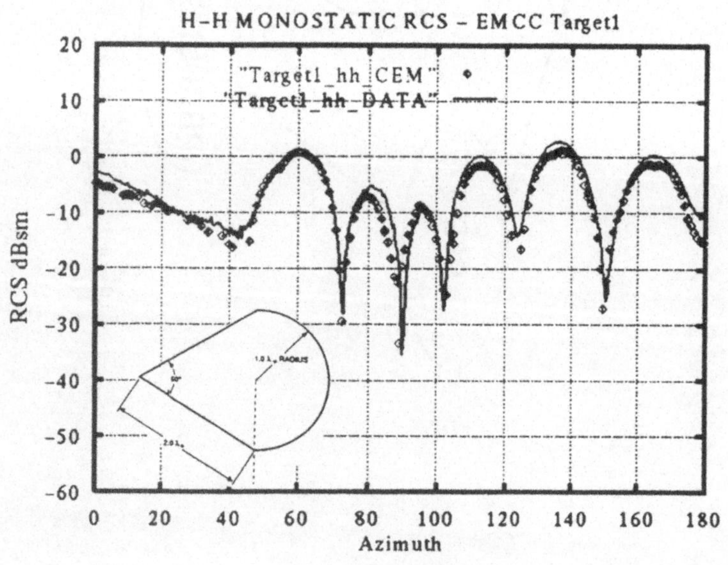

Fig. 2

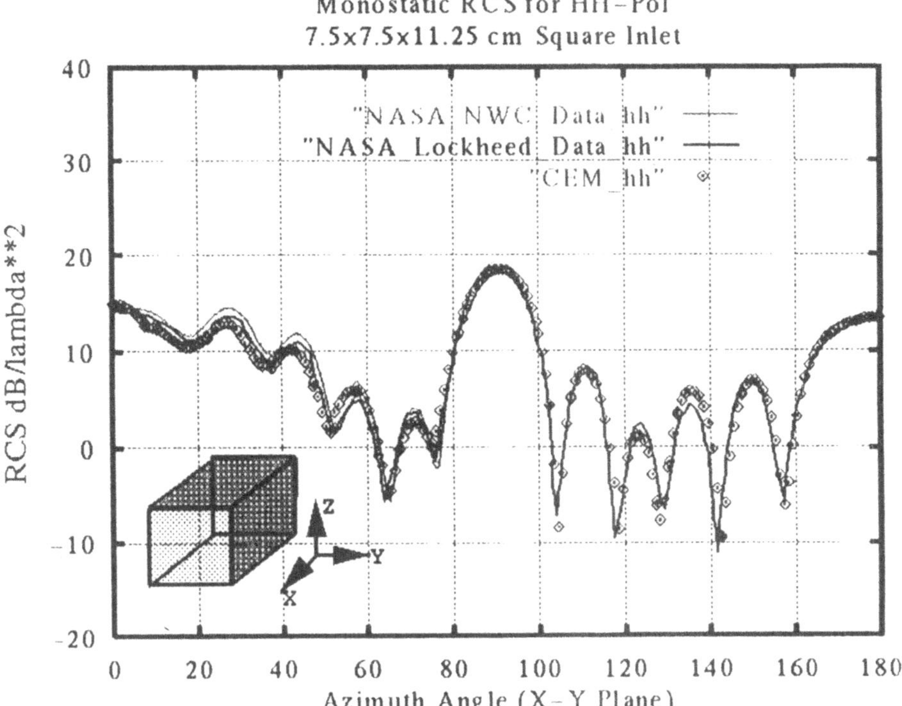

Fig. 3

Monostatic RCS for VV-Pol
7.5x7.5x11.25 cm Square Inlet
40
30
20
10
0
-10
-20
RCS dB/lambda**2
"NASA_NWC_Data_vv"
"NASA Lockheed Data_vv"
"CEM_vv"
0 20 40 60 80 100 120 140 160 180
Azimuth Angle (X-Y Plane)
Z
Y
X

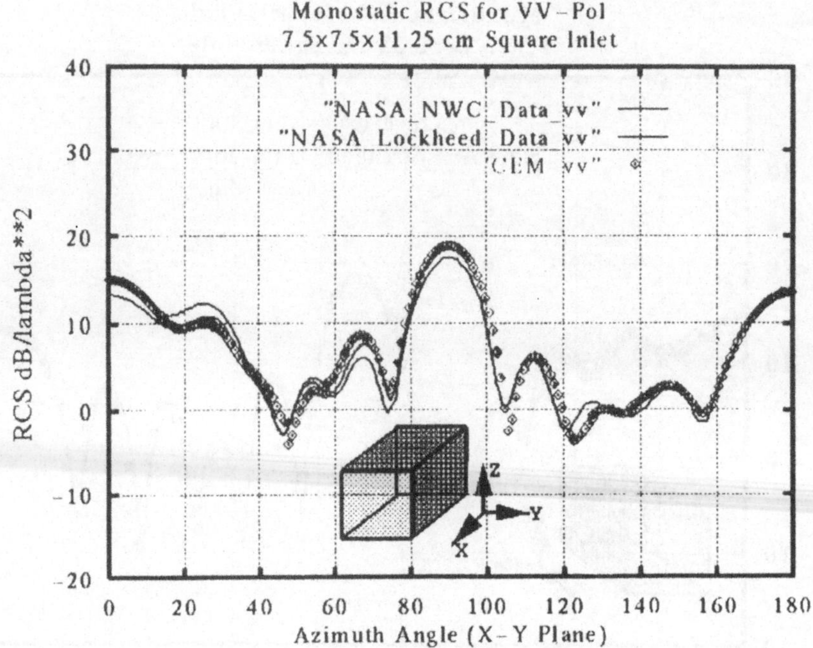

Fig. 4

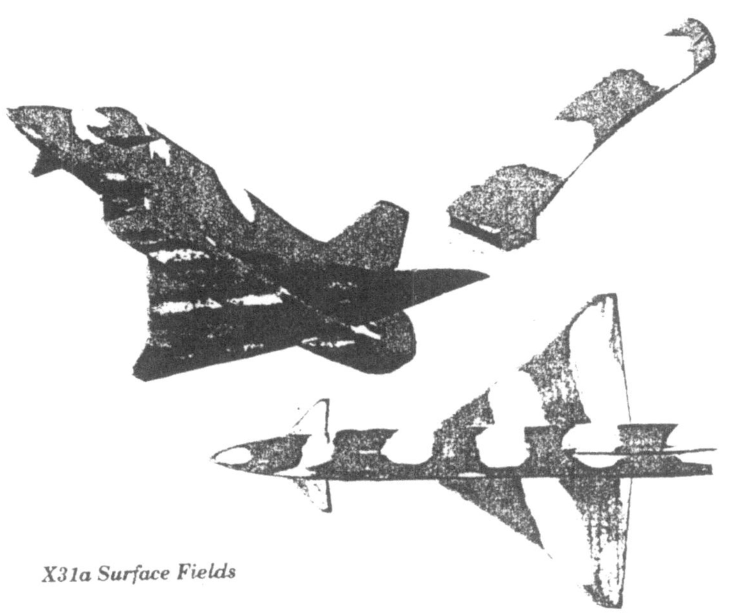

X31a Surface Fields

Fig. 5

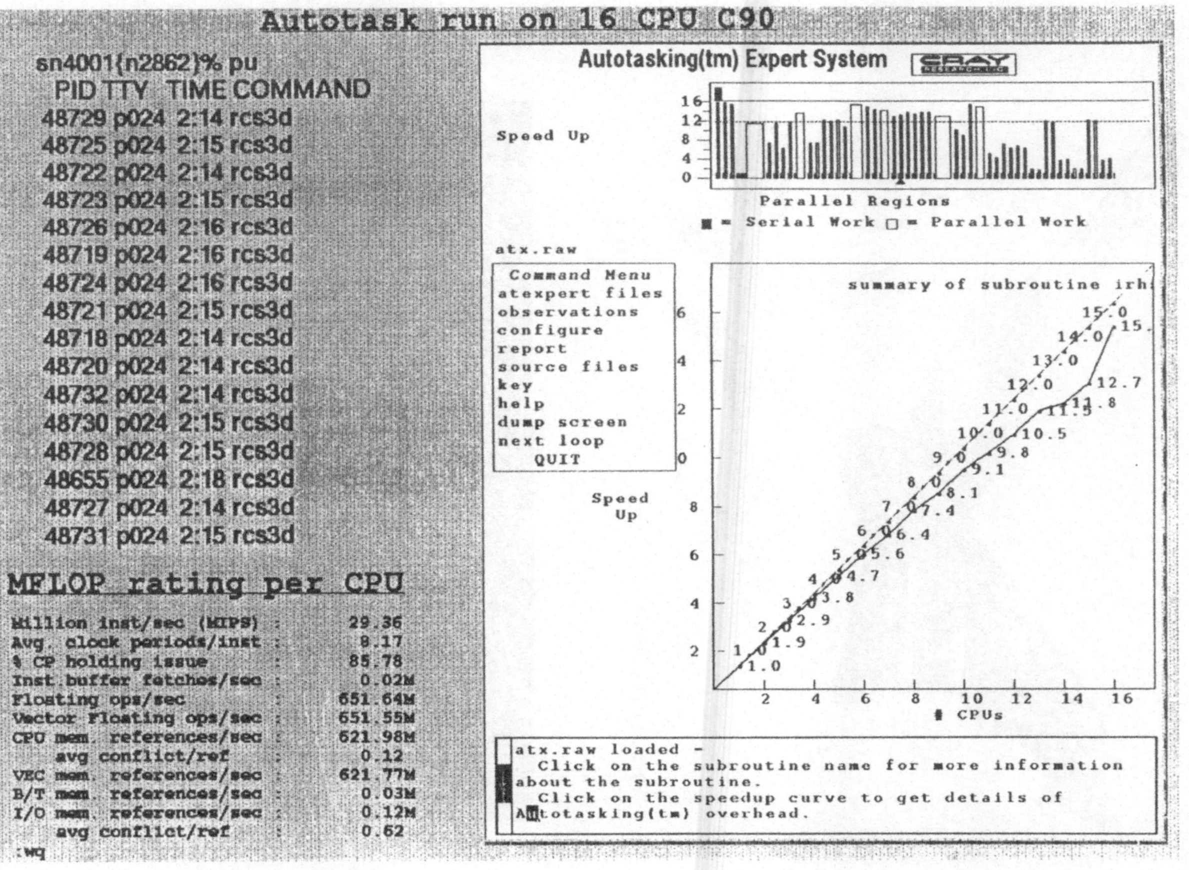

Autotask run on 16 CPU C90
sn4001{n2862}% pu
PID TTY TIME COMMAND
48729 p024 2:14 rcs3d
48725 p024 2:15 rcs3d
48722 p024 2:14 rcs3d
48723 p024 2:15 rcs3d
48726 p024 2:16 rcs3d
48719 p024 2:16 rcs3d
48724 p024 2:16 rcs3d
48721 p024 2:15 rcs3d
48718 p024 2:14 rcs3d
48720 p024 2:14 rcs3d
48732 p024 2:14 rcs3d
48730 p024 2:15 rcs3d
48728 p024 2:15 rcs3d
48655 p024 2:18 rcs3d
48727 p024 2:14 rcs3d
48731 p024 2:15 rcs3d
MFLOP rating per CPU
Million inst/sec (MIPS)   : 29.36
Avg. clock periods/inst   : 8.17
% CP holding issue   : 85.78
Inst buffer fetches/sec   : 0.02M
Floating ops/sec   : 651.64M
Vector Floating ops/sec   : 651.55M
CPU mem references/sec   : 621.98M
     avg conflict/ref   : 0.12
VEC mem references/sec   : 621.77M
B/T mem references/sec   : 0.03M
I/O mem references/sec   : 0.12M
     avg conflict/ref   : 0.62
:wq
Autotasking(tm) Expert System
CRAY RESEARCH LLC
Speed Up
16
12
8
4
0
Parallel Regions
■ = Serial Work □ = Parallel Work
atx.raw
Command Menu
atexpert files
observations
configure
report
source files
key
help
dump screen
next loop
   QUIT
Speed Up
summary of subroutine irh
15.0
14.0 15.
13.0
12.0 12.7
11.0 11.5 11.8
10.0 10.5
9.0 9.8
     9.1
8.0 8.1
7.0 7.4
6.0 6.4
5.0 5.6
4.0 4.7
3.0 3.8
2.0 2.9
1.0 1.9
1.0
CPUs
2 4 6 8 10 12 14 16
atx.raw loaded -
  Click on the subroutine name for more information
about the subroutine.
  Click on the speedup curve to get details of
Autotasking(tm) overhead.

Fig. 6

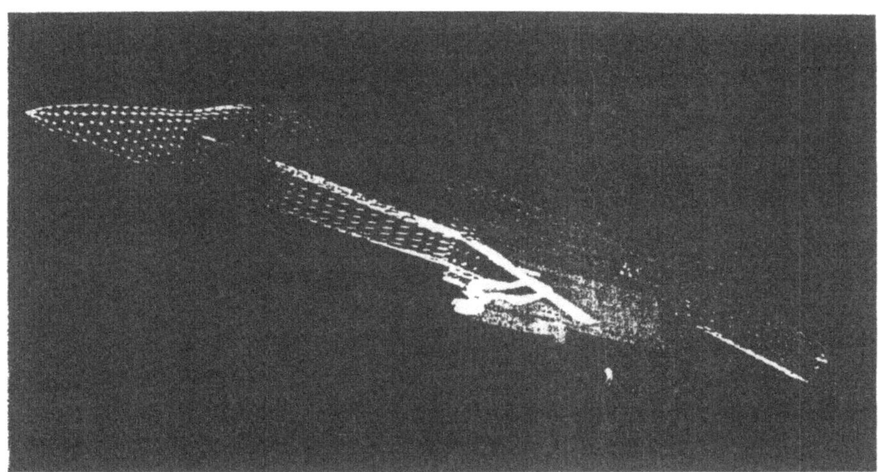

Fig. 7

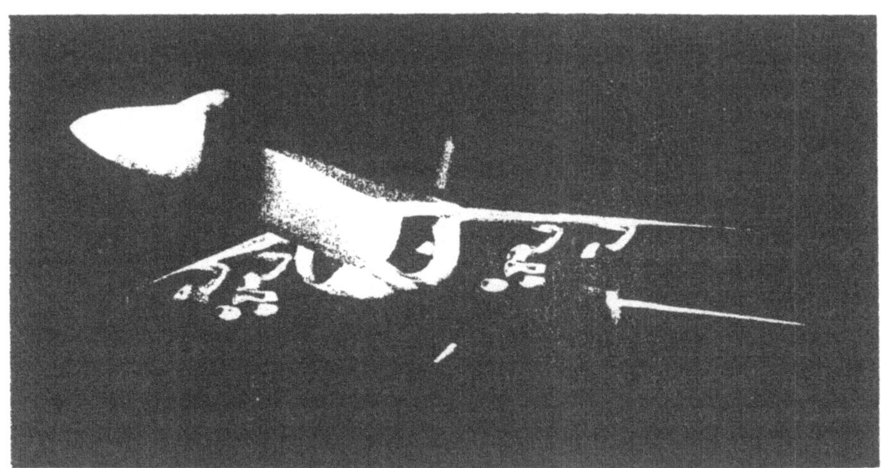

Fig. 8

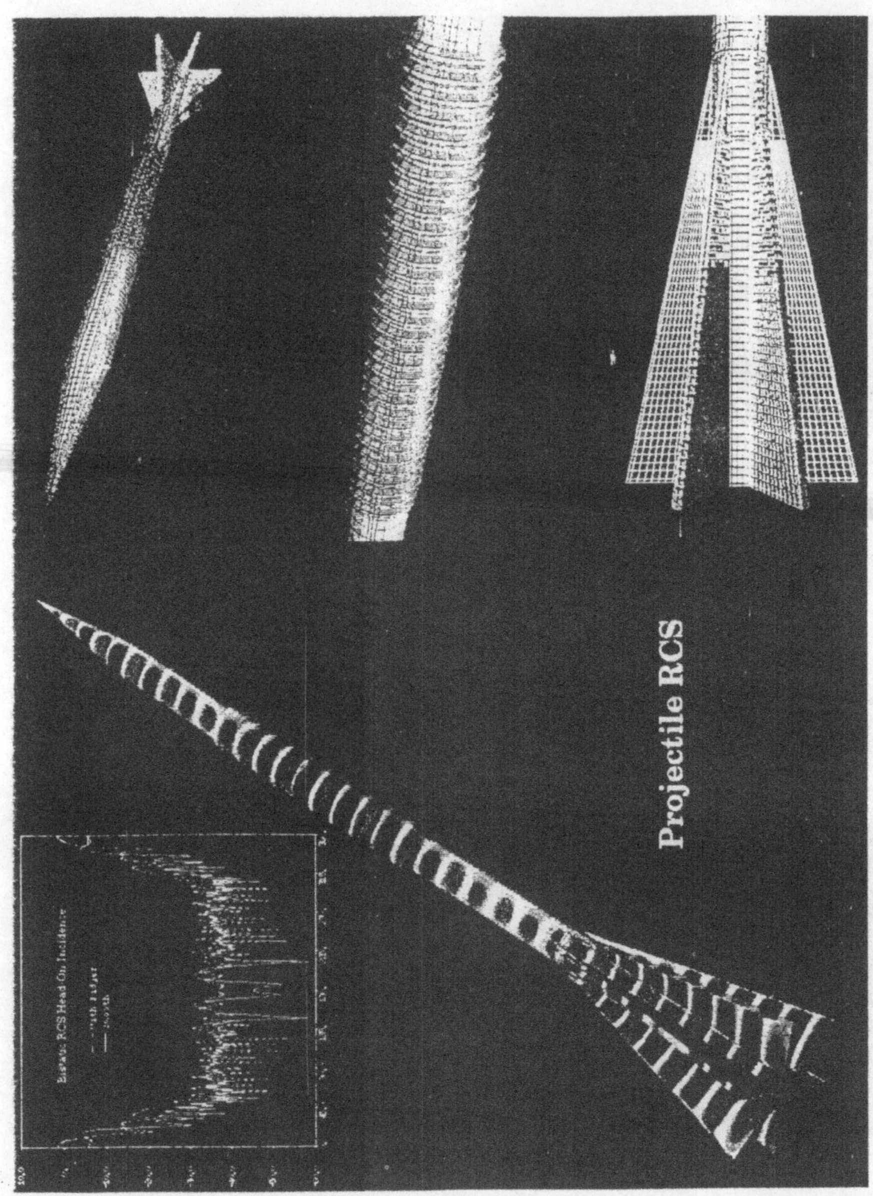

Fig. 9

Slotted Waveguide Antenna
Microstrip Patch Antenna
Low Pass Filter
CEM Applications

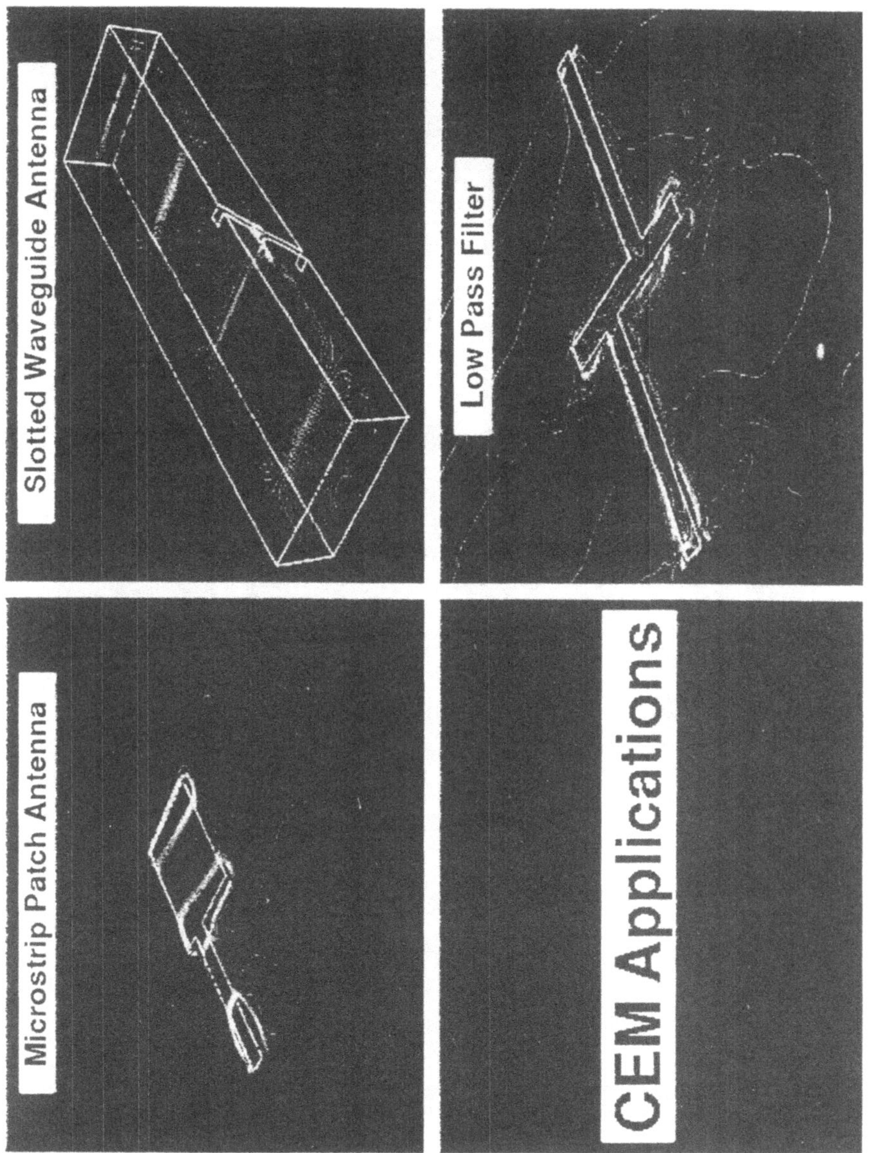

Fig. 10

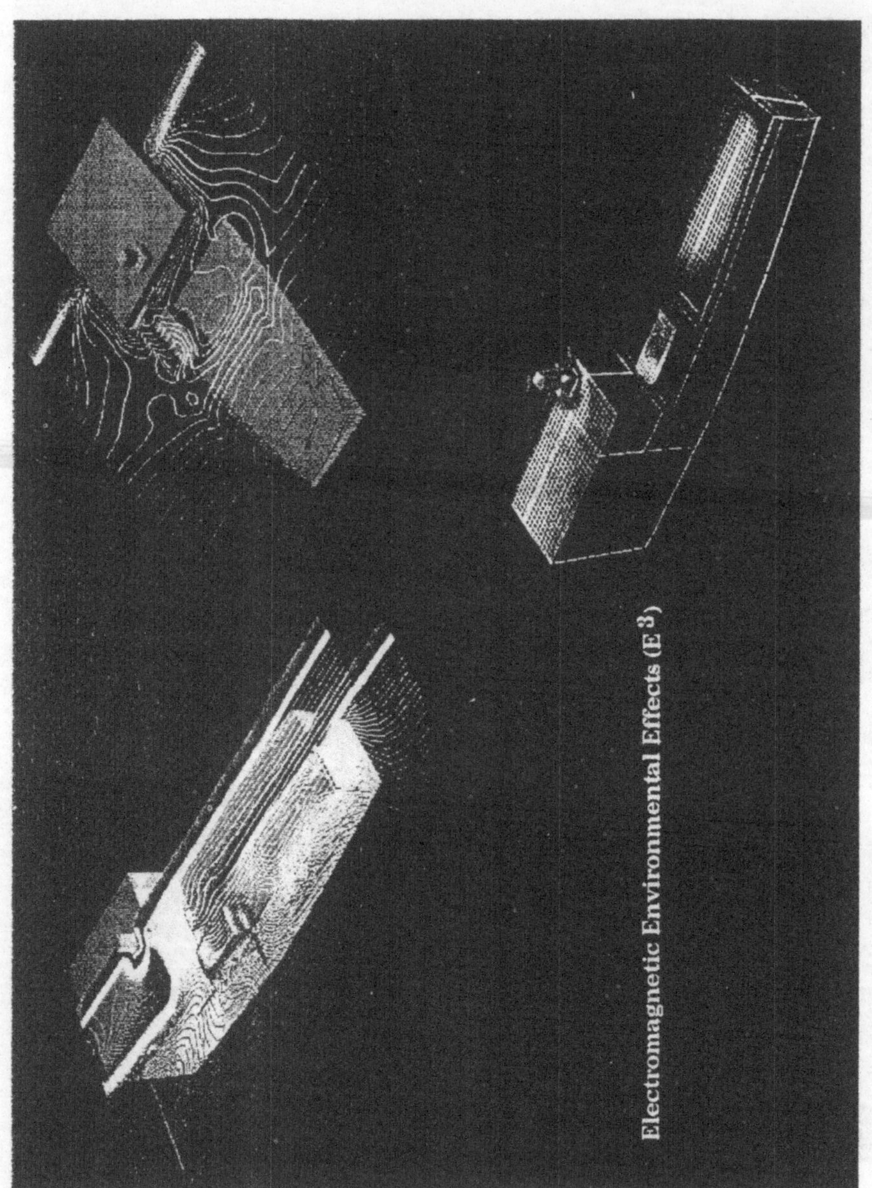

Fig. 11

N	0	1	2	3	4	5	6	7	8
# of Nodes	1	2	4	8	16	32	64	128	256
Grid Size	Elapsed Seconds								
$10 \times 10 \times 10$	2.87	1.605	0.903	0.505	0.350				
$20 \times 20 \times 20$	20.2	10.74	5.698	3.017	1.692	0.968♣	0.533	0.37	
$30 \times 30 \times 30$				9.21	5.27	3.08♠	1.746	0.994	0.588
$40 \times 40 \times 40$				20.76	11.04	5.975●	3.15	1.757	0.969♣
$60 \times 60 \times 60$							9.52	5.436	3.084♠
$80 \times 80 \times 80$							21.30	11.32	5.98●
$90 \times 90 \times 90$							31.65	17.35	9.46

Table 1. Demonstration of scalability with problem size

Function Spaces, Differential Operators and Nonlinear Analysis

The book contains invited surveys and shorter communications presented at the International Conference "Function Spaces, Differential Operators and Nonlinear Analysis", which took place in Friedrichroda (Thuringia, Germany), autumn 1992. The main subjects and contributors are:
- Linear and nonlinear elliptic and parabolic boundary value problems (H. Amann, P. Drábek, S. I. Pohožaev, T. Runst)
- Functional calculus and singular operators in spaces of Besov and Sobolev type (G. Bourdaud, A. Youssfi),
- Hardy type operators in weighted functions spaces (M. Krbec, B. Opižc, L. Pick, J. Rákosnik).

Edited by
Hans-Jürgen Schmeisser
and **Hans Triebel,**
both Jena

1993. 308 pages.
16,2 x 23,5 cm.
Bound DM 54,80
ÖS 428,– / SFr 54,80
ISBN 3-8154-2045-8

(TEUBNER-TEXTE zur Mathematik, Vol. 133)

B. G. Teubner Verlagsgesellschaft
Stuttgart · Leipzig

Jentsch/Tröltzsch

Problems and Methods in Mathematical Physics

This volume contains papers presented to the "10th Conference on Problems and Methods in Mathematical Physics" (10th TMP) held at the Technical University of Chemnitz-Zwickau, September 1993. The papers mainly focus on scientific computing in solid and fluid mechanics, problems with singularities, boundary integral methods, multilevel-finite-element methods, free boundary value problems, optimal control, and numerical analysis.
Prominent experts present excellent surveys, latest results and trends in their topics. Modern mathematical tools applied to solving problems in physics and engineering make this an important text for researchers at universities and in industry

Edited by Prof. Dr.
Lothar Jentsch and
Prof. Dr. **Fredi Tröltzsch,**
both Chemnitz

1994. 221 pages
with 18 figures.
16,2 x 23,5 cm.
Bound DM 44,80
ÖS 350,– / SFr 44,80
ISBN 3-8154-2058-X

(TEUBNER-TEXTE
zur Mathematik, Vol. 134)

B. G. Teubner Verlagsgesellschaft
Stuttgart · Leipzig

Hoschek / Dankwort
Parametric and Variational Design

Das parametrische Konstruieren hat in den letzten beiden Jahren erheblich an Bedeutung gewonnen: Ziel ist es, Konstruktionsvorgänge von Teilen, Werkzeugen, usw. mit Parametern zu belegen, so daß Änderungen der Konstruktion durch Verändern des entsprechenden Parameters erfolgen können, der Rechner wiederholt dann mit dem neuen Parameter den ganzen Konstruktionsvorgang. Bisher war bei einer solchen Änderung faktisch eine neue Konstruktion notwendig.
In dem Buch stellen Repräsentanten aus den Forschungszentren der (marktführenden) Systeme: Cisigraph, Computervision, IBM/Dassault, Intergraph, SDRC und Unigraphics ihren theoretischen Ansatz dar, ergänzt werden diese Berichte durch Übersichten zu dem Leistungsvermögen der beteiligten Systeme. Das Protokoll einer Podiumsdiskussion zwischen Anwendern und Entwicklern gibt einen Einblick in viele aktuelle Probleme aus dem Einsatz der CAD–Systeme

Prof. Dr.
Josef Hoschek,
TH Darmstadt
und Dr.
Werner Dankwort,
BMW AG, München

1994. II, 176 Seiten.
16,2 x 23,5 cm.
Paper. DM 54,–
ÖS 421,– / SFr 54,–
ISBN 3-519-02632-5

B. G. Teubner Stuttgart